T0332603

Being Mobile

Future Wireless Technologies and Applications

Do you need to get up-to-speed quickly on the technologies and services that could transform the wireless world over the coming decade? Whether you work directly with wireless or in a sector where wireless solutions could be beneficial (e.g. healthcare, transport, sensor networks, location and smart metering), this concise guide provides a critical insight into future developments.

For the first time, you will have a clear view of all the key technologies and all the sectors or applications in which they could be used, with a comparison of the positives and negatives of each technology and sector area. You'll also see where the technologies required overlap and so could bring benefits across multiple areas, as well as how the key drivers of change in the past may impact on the future.

Key technologies covered include

- mesh networks
- white-space/cognitive devices
- 4G/LTE
- femtocells
- software-defined radios
- enhancements to screens, keyboards and handset technology

Key sectors and applications discussed include

- wireless in the healthcare and transport sectors
- delivering social networking when mobile
- implementing the smart grid with wireless communications
- location and how it will lead to virtual reality
- wireless as an essential service and Government's desire for ubiquity

William Webb is Director of Technology Resources at Ofcom, the UK telecommunications regulator. He is also a Visiting Professor at Surrey University and DeMontfort University, and a Fellow of the Royal Academy of

Engineering, the IEEE and the IET, of which he is also a Vice President. He has previously published 10 books, 80 papers and 4 patents, and he is the Series Editor for the Cambridge Wireless Essentials book series published by Cambridge University Press. His industry experience spans a range of communication consultancies in the fields of hardware design, computer simulation, propagation modelling, spectrum management and strategy development.

Being Mobile

Future Wireless Technologies and Applications

WILLIAM WEBB
Ofcom

CAMBRIDGE
UNIVERSITY PRESS

University Printing House, Cambridge CB2 8BS, United Kingdom

One Liberty Plaza, 20th Floor, New York, NY 10006, USA

477 Williamstown Road, Port Melbourne, VIC 3207, Australia

314-321, 3rd Floor, Plot 3, Splendor Forum, Jasola District Centre, New Delhi - 110025, India

79 Anson Road, #06-04/06, Singapore 079906

Cambridge University Press is part of the University of Cambridge.

It furthers the University's mission by disseminating knowledge in the pursuit of education, learning and research at the highest international levels of excellence.

www.cambridge.org
Information on this title: www.cambridge.org/9781107000537

First published 2010

A catalogue record for this publication is available from the British Library

ISBN 978-1-107-00053-7 Hardback

To my family, who across the generations make use of mobile technology in quite different ways. To my mother and father-in-law, who make valiant attempts to use the technology even if they occasionally need 'IT support'; to my wife, who finds the Internet one of the greatest inventions of modern life; and to my daughters Katherine and Hannah, who see the mobile phone as a must-have means of conducting their social life. May the advances predicted here benefit you all in some way.

William Webb

Contents

About the author

WILLIAM WEBB

William joined Ofcom as Head of Research and Development and Senior Technologist in 2003. Here he manages a team providing technical advice and performing research across all areas of Ofcom's regulatory remit. He also leads some of the major reviews conducted by Ofcom, including the Spectrum Framework Review, the development of Spectrum Usage Rights and most recently cognitive or white-space policy. Previously, William worked for a range of communications consultancies in the UK in the fields of hardware design, computer simulation, propagation modelling, spectrum management and strategy development. William also spent three years based in Chicago providing strategic management across Motorola's entire communications portfolio.

William has published ten books, eighty papers, and four patents. He is a Visiting Professor at Surrey University and DeMontfort University and a Fellow of the Royal Academy of Engineering, the IEEE and the IET, of which he is a Vice President. His biography is included in multiple *Who's Who* publications around the world. William has a first-class honours degree in electronics, a PhD and an MBA.

In between working on telecoms regulation, his roles in institutions and writing books, William is a keen cyclist. He has ridden from Lands End to John O'Groats, done the Etape du Tour, in 2009 rode the first half of the Tour de France and in 2010 completed the Cent Cols Challenge – riding over 100 Alpine passes in just 10 days.

Abbreviations

3G	Third-generation cellular technologies
4G	Fourth-generation cellular technologies
AoA	Angle of arrival
A-GPS	Assisted GPS
A–D	Analogue-to-digital (convertor)
ADSL	Asynchronous digital subscriber line
AMR	Advanced multi-rate voice coder
C2C	Car to car
CCTV	Closed-circuit television
CDMA	Code-division multiple access
C/I	Carrier-to-interference ratio
DAB	Digital audio broadcasting
DECT	Digital European cordless telephone
DLNA	Digital Living Network Alliance
DSL	Digital subscriber line
DSP	Digital signal processor
DTT	Digital terrestrial television
DVB	Digital video broadcasting
DVB-H	Digital video broadcasting for handheld use
DVB-T	Digital video broadcasting for terrestrial deployment
DVD	Digital versatile disc
EDGE	Enhanced data rates for GSM evolution
EC	European Commission
FFT	Fast Fourier transform
FM	Frequency modulation
FPGA	Field programmable gate array

GEO	Geo-stationary orbit satellite
GPRS	General packet radio service (enhancement to GSM)
GPS	Global positioning system
GSM	Global system for mobile communications
GSM-R	GSM for railways
HAP	High-altitude platform
HD	High definition
HSDPA	High-speed downlink packet access
HSPA	High-speed packet access (for 3G cellular systems)
ICT	Information and communication technology
IMS	Intelligent multi-media service
IP	Internet protocol
IPTV	Internet-protocol television
ISI	Inter-symbol interference
ISP	Internet service provider
JPEG	Joint Pictures Experts Group
LED	Light-emitting diode
LEO	Low-Earth-orbit satellite
LTE	Long-term evolution (of cellular networks)
$MtCO_2e$	Mega-tons of carbon dioxide or equivalent
MEMS	Micro-electronic mechanical system
MIMO	Multiple-input multiple-output (antenna system)
MPEG	Motion Pictures Expert Group
MVNO	Mobile virtual network operator
OFDM	Orthogonal frequency-division multiplexing
OLED	Organic light-emitting diode
PDA	Personal digital assistant
PLT	Power-line telecommunications
PVR	Personal video recorder
R2V	Roadside to vehicle
RF	Radio frequency
RFID	Radio-frequency identifier
SD	Standard definition
SDR	Software-defined radio
SIM	Subscriber-identity module
SMS	Short-message service
TDoA	Time difference of arrival
UHF	Ultra-high frequency

UWB	Ultra-wideband
VoIP	Voice over IP
VSAT	Very-small-aperture terminal (for satellite transmission)
W-CDMA	Wideband code-division multiple access

1

Introduction

1.1 The rationale for this book

Wireless communications seems to be an area of frequent and rapid change. New concepts such as updating a Twitter account from a mobile phone arise and become pervasive in less than a year. New devices like the iPhone capture the public imagination within weeks of being launched and in turn change the relationships between the key players in the industry. Satellite navigation seems to be rapidly incorporated into most mobile devices, which themselves are typically replaced within 18 months. Compared with most other industries and consumer products the rate of change is startling. Even in other industries such as the automotive industry, some of the new features such as adaptive cruise control, advanced satellite navigation and collision-control radars are due to advances in wireless technology.

Understanding what is on the 'wireless horizon' – namely what developments are now being considered, developed or trialled – can help make sense of how the wireless world is likely to evolve. This book is about scanning that horizon, identifying the important developments and discussing how they will impact on the world of wireless communications over the next decade or so.

As will be seen in the chapters of this book, simply identifying interesting new technologies is far from sufficient. There have been many 'interesting' new wireless technologies that have failed to live up to their initial promise – mesh wireless networking is one – for a variety of reasons, many of which are not technical. These include the structure of the market, the cost of their provision, the complexity they might entail and, not least, whether they solve a problem that cannot be addressed more readily by other means.

Hence the structure of this book has been divided into two parts. The first looks at new technologies that are 'bubbling'. These are ideas that are being discussed around the industry, perhaps in academia, research laboratories or conferences. Because the proponents of each idea tend to 'hype' their invention in the hope of getting it widely adopted they are generally relatively easy to spot. Some have been around for many years but have yet to be implemented widely; others are relatively new. The second part of the book looks at application areas such as transport and entertainment and considers the requirements of those sectors. This allows a possible matching between the capabilities of emerging technologies and the needs of particular sectors as well as a discussion of the non-technical barriers that might prevent implementation.

The use of wireless covers an enormous range of activities. This book is mostly concerned with those uses of wireless that affect consumers and change everyday life.

One of the key areas it covers is the cellular phone. This is a large industry involving manufacturers both of infrastructure equipment and of handsets, operators who run the networks and provide service to end users and increasingly applications providers. Of course, there are many associated companies providing a wide range of inputs from chipsets to consultancy and from billing to mast construction.

Another area of increasing importance is the unlicensed use of wireless such as WiFi, Bluetooth and the myriad devices in the home such as garage-door openers. The industry here tends to involve predominantly the manufacturer of the device, although, with wireless embedded in a wide range of goods, the range of manufacturers is equally large.

Wireless is also becoming embedded in many of our possessions including vehicles, computers, toys and so on. All of these manufacturers are affected, and affect, the wireless industry. For example, as we will see, the ability to use wireless in the car changes some of the ways in which it is designed and features that it offers.

This book does not consider wireless in a number of less consumer-focused applications. It does not cover wireless in the military, except where technologies pioneered by the military are now making their way into the commercial world. It does not say much about satellite systems, partly because there is little to say about these (other than that they are steadily improving). Fixed links (or microwave point-to-point links) are not covered, again partly because there is little change expected here and because they are also not direct to the end user. Finally, one of the biggest users of the radio spectrum, radar systems, also does not get a mention, for very similar reasons to the other categories. Actually, with radar systems there is the possibility of using novel designs that would dramatically reduce the use of spectrum without compromising performance, but the speed at which new radars are introduced is so slow and the need for safety case verification so great that the pace of change is glacial.

This book comes at a time when the wireless industry is facing new challenges. For most of the last 25 years wireless has been a dynamic growth industry with companies reporting ever increasing profits and with subscriber numbers rising dramatically. All that has now changed and many companies see declining revenues. Manufacturers have gone into bankruptcy or merged and operators are increasingly looking to mergers to save money. Operators are also facing a challenge from data volumes that have grown dramatically but without much associated increase in revenue. The industry is seeking a technical solution to this problem, but it is more likely that the solution will come from changed value chains, use of smaller cells and an overall change to the landscape of the wireless industry. These issues will be addressed at appropriate points within this book.

1.2 Looking through the rear-view mirror

Why look backwards?

Before looking forwards it is informative to look backwards and consider those technologies and applications which have had the greatest impact on wireless communications over the last 10–20 years. This may help us to understand the type of developments that have changed the world and, while the past is not always a good predictor of the future, there are always lessons that can be learnt.

As will be explained, many developments are interlinked. Some make use of the same underlying advances in technology; others are enabled as a result of another development and sometimes form virtuous or self-reinforcing circles. Hence, discussing developments in a logical or chronological fashion is not easy and a different approach is adopted here, namely that of looking at the most important developments and then examining their implications.

Key changes in communications

Arguably the biggest development over the last 10–15 years has been the emergence of the Internet as a core component of our everyday life. The Internet has resulted in new models of business, new ways of working, an extraordinary increase in access to information and rapid innovation across the entire spectrum of communications. The original enablers of the Internet are well chronicled and include the work on Arpanet and the development of hypertext and other protocols. From a communications viewpoint, the advent of broadband communications to the home did much to make the Internet a more valuable tool. The broadband revolution itself was broadly enabled by DSL technology, which became possible as a result of improving processing power within silicon. Widespread access to the Internet on highly standardised platforms then enabled applications such as Facebook to be written in just one format and rapidly adopted. Heavy reliance on the Internet has in turn made it increasingly important to be able to access it even when mobile, driving take-up of mobile data, WiFi connectivity and browsers on mobile phones. We may still be only at the start of this revolution.

Another key development has been the improvement in mobile devices, as epitomised by the iPhone. Part of this has been improvements in the user interface, including touchscreens and more intelligent software, but underlying the change are improvements in component technology, including miniaturisation of cameras, massive increases in memory size and improvements in chip fabrication, enabling multiple wireless technologies to be incorporated into the same device. The iPhone in turn drove changed business models including the success of the applications store. Such concepts had been tried before by mobile operators but failed because they required a very wide range of devices with many different operating systems to be supported and because the user interface often made it difficult to use the applications. The iPhone changed this by offering a single platform with a flexible user interface.

Applications in turn are driving a change in the idea of what a phone is for, leading to changes in the connectivity required. For example, it would have been hard to envisage five years ago phones whose design was centred around updating of social-networking sites. This further increases the attractiveness of wireless connectivity such as WiFi, making it more ubiquitous, making chipsets less expensive and powering another virtuous circle.

Improvements in storage have also driven key developments in entertainment. The CD has increasingly been displaced by MP3 players as their functionality has improved. Of course, this in turn was accelerated by iTunes, which linked device developments with content and helped pave the way for the iPhone. Hard-disc storage has also displaced the video-cassette recorder with the PVR as a mechanism for recording and time-shifting programmes and even the DVD is now under threat as hard discs become large enough to hold archive video material. It is clear that delivery of video content will change over the coming years as these factors play out.

As a separate development, the availability of GPS has driven widespread adoption of satellite navigation and laid the basis for many location-based services.

Technical drivers

The technical drivers that underlie these changes – those that we might have hoped to have predicted 15 years ago – are predominantly related to storage, processing power, lower-cost and denser chips and better displays and input devices.

Our ability to store data continues to grow by around an order of magnitude every six years. This applies both to hard-disc storage on PCs and in PVRs and to flash-memory storage in portable devices. To put this into perspective, about every 18 years there is a thousand-fold increase in the amount of storage available. This trend has been visible since the mid 1970s and has remained fairly consistent, with slight periods of acceleration and deceleration away from the trend. Predicting this change in storage was relatively easy and some of the implications, such as being able to store an entire music library on a portable device, were known many years in advance. There is no sign that this trend is slowing or reaching any fundamental limits, as will be discussed further in later chapters.

Processing power is a similar story. Moore's Law goes back even further and has been followed with uncanny accuracy over many decades (most probably because it has become a self-fulfilling prophecy). This drives not only enhancements in processors but also increasingly dense integration on chipsets. The result has been that concepts that were overly complex at one point become implementable later. (As an aside, when GSM was standardised in the late 1980s, it was known at the time that it was not possible to implement a portable device with the available technology, but the assumption was that this technology would improve sufficiently to allow portable devices – and it did.) However, as regards wireless, we may be coming towards the end of this trend, more because the battery power available cannot support more powerful processors than because it is not possible to manufacture them. Just as with storage, increased processing power was a very predictable trend.

Better displays and in particular touchscreen displays were more difficult to predict. There are no well-known 'laws' of evolution in display technology, instead there tend to be breakthroughs as new materials are brought into use or new manufacturing approaches enable cost-effective displays. How displays might change in the future is likely to be as much of a key driver as it has been in the past and this is considered further in Chapter 9.

Application drivers

While technology can enable new devices, it is the applications they can be used for that result in consumer interest and adoption. As mentioned above, one of the key drivers has been the Internet, which has enabled a wide range of applications including communications of all sorts, on-line purchasing, targeted advertising, information search, information sharing and a range of business connectivity functions. Many of these are things that consumers were already doing (communicating, purchasing, etc.) but could now be performed more effectively.

On-line purchasing was one of the early Internet applications leading to Amazon and a wave of Internet retailers. At first take-up was slow as consumers persisted in the 'old way' of doing things and because they harboured doubts and concerns about security, payment mechanisms and delivery issues. However, as confidence grew and early adopters told others of their good experiences, adoption grew quickly and moved from the most obvious areas such as book purchase where actually seeing or holding the product was of little relevance to areas such as clothes retailing and grocery shopping. After initially replicating the physical shopping experience, a range of novel services such as 'if you liked that you'll probably like this too' and consumer ratings were added, further increasing the value to end users. Comparison sites sprang up, enabling users to compare prices across a range of vendors. So on-line purchasing started by replicating an existing experience and then evolved to add new services and value.

On-line search was another obvious function to add to the Internet. Initially appearing simply because there was no other way to access the mass of data available on-line, search improved rapidly as clever algorithms were devised to present increasingly useful results. This is an on-going process as researchers seek algorithms that will 'answer questions' rather than just look for matches to search strings. Search broadly replaced either going to a library or asking around a number of colleagues or organisations for access to information. Search often has a similar function to advertising – with search a consumer may be trying to discover products, whereas with advertising a company may be trying to inform consumers about their products. Thus it was an obvious step to link the two together – forming the business model that has driven Google's success. This is now driving fundamental changes in advertising, with spending shifting from traditional media to the Internet in a way that is having knock-on consequences for newspapers, broadcasting and other entities with advertiser-funded revenue models.

Search, advertising and on-line purchasing have fed off one another. Search makes it simpler to compare prices across multiple stores and discover information about products than it would be via 'conventional' shopping. This in turn feeds more on-line purchasing, increasing the value of search and click-through advertising. Internet retailing is often cheaper for the retailer, who no longer needs a physical presence, further increasing the overall attractiveness. With all these self-reinforcing characteristics it is unsurprising that on-line purchasing initially grew slowly but then reached a 'tipping point' where growth accelerated sharply.

However, this evolution is threatening to lessen the value of the original activities. So, for example, search results are becoming less about what can be 'found' and more about what has been paid for by advertisers to be shown to those searching for particular items. In some cases there may be no conflict between these and in others search engines such as Google appear to differentiate between paid-for responses and what has been found. Others try to increase the likelihood of their site being listed by a search engine by modifying parameters of the site to make it more likely that it will be found – tending to favour those with the greatest resources. Perhaps this behaviour might be compared with colonising the West in the USA. The Internet to date has been somewhat like Frontierland with little in the way of rules and much in the way of land-grabs. Once rules start to emerge, then 'Main Street' starts to make its presence felt, increasing the dominance of large organisations and decreasing that of individuals. Quite where this will end is unclear – perhaps with alternative types of search deliberately trying to throw up unexpected and unusual results.

Probably the largest and least predicted growth area has been social networking. This started with sites such as Friends Reunited and YouTube but really took off with the advent of Facebook and MySpace. It found yet another direction with Twitter. Unlike retailing and search, these were applications with little parallel in the pre-Internet world, making them much harder to predict. Many appear to have emerged and become highly successful in very short periods of time, partly because of the ease with which applications can be downloaded and people become members, often without needing to spend any money. At the moment there appears to be no end in sight to the social-networking phenomena, although some are wondering whether users will tire of constantly updating their Facebook entry or reading multiple tweets.

Most of these applications proved themselves on fixed networks, initially on PCs connected by wires and low-speed modems to the Internet, but, as the applications proved their value in a fixed environment, users started to want to access them when mobile as well. Mobile search is especially valuable as it can be linked to location. Accessing social-networking sites from a mobile is seen as increasingly important, with some mobiles being designed around this functionality. On-line purchasing is heading in new directions with wireless – for example the ability to wirelessly purchase an e-book for download to a reader. Computers are increasingly wireless, with WiFi providing the final connection for many even if the main connection to the building is via wires. So understanding the applications that are successful on wired systems may provide some guidance as to applications that are likely to expand to wireless – fixed search expanded slowly but steadily to mobile search. However, not all fixed applications go mobile, some take many years if not decades to make the transition and mobility brings additional functionality that can change some applications.

Linking the two

Apparent from reading the section on technologies and then that on applications is how unrelated they appear. Understanding the technology drivers would appear to be of little use in determining which applications would emerge and, equally, looking at the key drivers for the applications reveals little about the technological changes needed. That explains the reason for looking at both in this book rather than trying to predict the future from the evolution of technology or an idea for a new application.

Many of the applications discussed had long been forecast – for example the dot-com bubble of 2000 was based on the idea that on-line retailing would change the way that we shop. In retrospect, many of the forecasts were broadly correct in their magnitude but optimistic in their timing. Some of the discussion above about the virtuous spiral of growth of inter-related services suggests why this might be. Applications are based on a very complex inter-related 'eco-system' that includes manufacturers, operators, other service providers and, importantly, the early adopters and advocates amongst the end users. For example, for location-based services manufacturers need to build easy-to-use GPS location systems within the handset, operators need to provide a framework for location-based services, entities like mapping companies need to produce appropriate offerings, Google needs to provide location-enabled search and early adopters and key influencers need to be enthusiastic about the service in order to convince others to adopt it. The relationships among all these players are complex, with some positive- and some negative-feedback loops. Models of such situations show how dramatically different outcomes can be achieved with relatively little change in inputs, and 'tipping points' are often observed. The complexity is not helped by the tendency of those in industry to look optimistically at the services they are working on, and for analysts to prefer reports with positive rather than negative outcomes.

The existence of tipping points – values of particular input variables at which the prediction of the model suddenly shifts from no growth to the hockey-stick – makes it almost impossible to predict accurately the success of such services. The chances are that most will be predicted to be successful for many years during which they will languish and then suddenly, for reasons that might not even be apparent, or appear of little relevance, they will take off rapidly. All we can do is learn from the models as to what behaviours would be most likely to result in success; but actually we know this already – to be successful all elements of the service launch must be near-perfect. The technology must work, the service must be easy to use, the pricing must be attractive and the marketing must attract the right early adopters, who must be deeply impressed. If any one element is not quite right it could be enough to prevent the service succeeding. That much is common sense. The difficulty, as always, is for all the companies to work together in a way that is competitive but collaborative, and that embraces standards but allows competitive differentiation. This is very hard – the incentives for individual organisations are rarely such that they work together well. What tends to happen is that individual elements slowly get solved and, when the last one falls into place, the service takes off.

1.3 Learning from previous predictions

The previous section looked back at the most important developments in communications and related areas in the last 10–15 years. That is relatively easy to do with hindsight, but might we have predicted these developments had we been writing this book in 1995? The only way to really tell is to look at some predictions made in the past and ask how accurate they were. Actually, this turns out to be somewhat difficult to do. There are many sources of such predictions and they are all somewhat different, so a different result will be achieved depending on the source.

Even more difficult is that many predictions make use of scenarios. Scenarios, on the face of it, represent a very sensible approach to forecasting. There are some variables that appear just too difficult to predict, such as whether mobile TV will take off. It seems better, then, to model a range of scenarios, often representing extreme cases. This would work well if all the scenarios, or almost all, pointed to a similar outcome. For example, if 4G were needed under all reasonable scenarios, then the analysis would have demonstrated strongly that its emergence was nearly certain. But in the communications sector this very rarely happens. Instead, the 'status quo' scenario shows that no new networks or technologies are needed while the 'wireless-data explodes' scenario shows that networks will need a sixty-fold increase in capacity or more. Effectively, scenarios demonstrate that in order to make a bet on the future you have to pick a particular scenario. Sadly, the forecasters who use scenarios rarely do. They simply present their set of scenarios and leave it for the reader to select their preferred one.

So in this section we discuss a set of predictions that we made in 2000 when we authored a book entitled *The Future of Wireless Communications* [1]. After a detailed look at the key drivers, the book made specific predictions as to how the world of wireless communications would look in 2005, 2010, 2015 and 2020.

We published another book in 2006, this time entitled *Wireless Communications: The Future* [2]. This also made specific predictions about 2011, 2016, etc. The predictions in these books were made taking a range of factors into account, including network modelling, an understanding of trends such as 'Moore's Law', a look back of up to 20 years to understand the pace of change, assessment of standards currently being developed and input from six 'gurus' as to their views and predictions from experts in specific areas such as screen technology.

In 2000 we predicted that by 2010 homes would become wireless hotspots, with dedicated home networks probably using the Bluetooth technology. Most homes would be connected via broadband with data rates in the region of 10 Mbits/s. Mobile phones, or by that time 'communicator devices', with a wide range of functionality would work with these home networks. Allowing phones to work in the home and office would bring a need for work–life differentiation that would be solved by re-direction functions in the network that reformatted, forwarded or sent incoming messages to message boxes according to circumstances and preferences. We thought that video communications would start to become used, making up about 2% of calls.

We predicted that people would communicate more readily with machines using speech recognition and that machines would increasingly communicate with each other. Networks would have packet-based cores and we thought that public WiFi hot spots would become commonplace.

Most of these appear about right. Homes have deployed wireless nodes, albeit WiFi rather than Bluetooth. Handsets can work on in-home WiFi networks, although this functionality is not yet widely deployed, more because it has not been in the interest of cellular operators to see this happen rather than due to technical reasons. Broadband connections are available to homes and data-rate requirements are somewhere in the region of 10 Mbits/s, although they are not always met. IP cores are well established in cellular networks and public W-LAN hotspots are ubiquitous in buildings in cities.

Some predictions proved over-optimistic. 'Re-direction functions' whereby messages would be intelligently re-routed have not occurred because they have proven overly complex; instead users have made use of different communications mechanisms to handle different message priorities. The prediction that speech recognition would be 'solved' by 2010 and would become the way in which people interacted with machines, rather than using keyboards, appeared entirely plausible in 2000 given the rapid progress that had been made in the late 1990s. However, this progress slowed and it proved very difficult to get accuracy up from around the 98% mark to the near-100% level needed to make it useful. Speech recognition is still making slow progress and is finding more applications, especially where the vocabulary it needs to recognise can be limited – for example, when checking a train timetable – but it may be another 10 years before it becomes widespread and trusted.

Video communications is more widely used, from the basic (but free) Skype services to the advanced video-conferencing suites pioneered by Cisco. However, it has proven to be of very little interest to mobile-phone users. Nevertheless, the basic premise that video communications would grow but would still be a very minor part of overall communications was about right.

While these specific predictions do not seem far off, and certainly suggest that predictions 10 years into the future are possible, more telling was a 'day in the life' vignette provided to illustrate what life in 2010 might look like. The vision saw a user (John) being woken by his communicator, which had already pre-programmed the home heating and downloaded journey details to his car satnav. John had personalised news delivered to his car audio and his emails were automatically read to him as he drove – none of these are widely available in 2010. Other predictions were more accurate, with John downloading emails from an airport hotspot, having only limited email facilities when flying and being able to connect to a hotel hotspot in the evening.

In 2006 we predicted that fixed networks would be steadily enhanced with fibre slowly deployed and upgrades to asynchronous digital subscriber line plus (ADSL2+) occurring in selected areas. We thought that broadcast TV would have changed little, with only small amounts of Internet TV viewing and little mobile TV access. In cellular networks we said that 3G would have become the most widely used cellular network, offering data rates up to 1 Mbits/s, while WiMax would have had no impact. We predicted that mobile Internet browsing would increase significantly.

We predicted that short-range wireless would continue to be dominated by WiFi and Bluetooth and that hotspot numbers would grow. In the home we felt that some security systems would start to use the home WiFi network by 2011. Handsets would continue to add increased functionality, with video storage and more advanced keypads and touchscreens. We predicted that GPS would be fitted only to smartphones but WiFi in handsets would become widespread. We thought that about 20% of users would have a converged handset for home and wide-area use but no real convergence between broadcast media and telecoms would have happened. In terms of industry structure we thought that a few service providers would start to emerge by 2011 but would make slow progress.

In predicting the user experience we said that convergence would occur between the home and cellphone and wireless home networks would increasingly lead to home automation. Users would make their first videocalls and the phone would become multi-purpose, displacing other specific devices.

There is something of a pattern emerging here. The specific predictions are all broadly correct. Those for fixed, broadcast and mobile networks are all accurate. The predictions for short-range wireless are also accurate, as were those for handsets. However, the predictions for video-calling and convergence look rather optimistic, broadly for the reasons discussed earlier. Service providers in the form of mobile virtual-network operators (MVNOs) are steadily emerging and do look to be having an increasing impact.

However, as with the 'day in the life' prediction, the predictions of user experience made only four years ago appear optimistic. With little convergence between home, office and wide-area use, filtering solutions have not been needed and wireless home control applications have been slow to emerge. Not all was incorrect though, with higher data rates, video-calling capabilities and better devices being correctly identified.

Some would suggest that predicting five years out into the future of wireless communications is pointless because the world changes too quickly and too unpredictably for such predictions to stand any chance of being correct. The evidence here is that this is not true – the predictions are generally accurate. Indeed, the first point of note from comparing the predictions with the reality is that the technology and networks that underlie the world of wireless communications change very slowly indeed. Where predictions were in error it was always due to predicting that change would happen more quickly than actually transpired.

The second point of note is that the technological predictions were generally the most accurate. Predictions of data rates and device capabilities were rarely in error. The spread of technology, in terms of the number of devices in homes, chipsets in handsets and so on, was also generally predicted accurately. With many well-known 'laws' of progression in areas such as storage and with the ongoing standards work providing a good indication of future consumer devices, this area of prediction is relatively easy.

The final point is that predictions of the services or functionality available to end users were almost always too optimistic. Home automation of any sort appears stubbornly slow, although perhaps energy-efficiency concerns might drive some changes in the future. Even more in error were predictions concerning 'personalised' services in which mobile devices make intelligent decisions on behalf of their owners, covering features from message processing to automatic re-routing of travel plans. Further understanding of the difficulties involved now suggests these may never emerge, with users instead finding specific solutions to specific needs, such as applications or widgets that perform functions like accessing train timetables. The safest bet when predicting the future of wireless, it would appear, is to expect little.

1.4 The organisation of this book

This book is divided into two parts, the first considering new or promising technologies and the second looking at application areas. For the most part this distinction is clear, although there are some important issues, such as environmental concerns, that fit neither category cleanly.

The technology section covers a range of different concepts including the following.

- The role that 4G cellular systems might play in communications networks of the future.
- Small home cells, or femtocells, and whether they will succeed and change the architecture and business model of cellular operators.
- Satellites and high-altitude platforms and the possible role they might play.
- A look at where mesh networks can bring advantages.
- An assessment of whether moving to a software-defined architecture is likely and will bring changes.
- A discussion of cognitive or white-space devices, assessing what applications they might be used for.
- A look at whether voice and video codecs are likely to improve over time, reducing the data rates needed for transmission.
- Consideration of the various key components in devices such as screens and batteries and the likelihood of significant advances in each.
- An examination of the different possible architectures for wireless networks, including a look at the role of ever-smaller cells.
- Finally, a discussion of whether environmental concerns will have any impact on the provision and use of wireless networks.

The solutions section then looks at a number of areas where wireless technologies might be applied, or which might drive the direction of the evolution of wireless systems.

- A look at the user, their needs and relevant trends.
- The use of sensors in a range of different networks and applications.
- Social communications including social networking as a key driver for wireless growth.
- The growing use of location information and the applications that might emerge as a result.
- The role of wireless in healthcare, including assisted living, transport and the entertainment sectors.

- The emerging discussion around the smart grid as a way of better providing energy and the implications this might have.
- The growing trend towards a universal service requirement for wireless connectivity and the role of the regulator in the future of wireless.

The book finishes with a summary chapter, which does not set out to predict any single future but rather brings together the key issues and conclusions from each of the separate chapters.

References

[1] W. Webb, *The Future of Wireless Communications*, Artech House, 2001.
[2] W. Webb, *Wireless Communications: The Future*, Wiley, 2007.

PART I TECHNOLOGIES

2
4G cellular

2.1 Introduction

We start our look at wireless technology with what are, to date, the most important wireless systems – the cellular communications networks. They are the most important because they create by far the greatest value from the use of wireless communications to date [1] (although some suggest that the role of short-range unlicensed communications will become increasingly valuable over the next decade).

Cellular communications has been one of the great success stories for wireless since the early 1990s. The number of subscribers, networks and mobile phones has grown very quickly and the networks have gone through three major generations of technology. However, this is now changing and network operators face the following key challenges.

- Growth in subscriber numbers is mostly at an end, except in developing countries, and for the first time annual revenue has started falling as competition has reduced prices.
- Traffic levels across wireless networks are growing rapidly as mobile broadband becomes popular, causing network congestion but not resulting in the increased revenue needed to invest in additional resources.
- Services and applications are increasingly being provided by others, sidelining the operators, who are becoming 'bit pipes', reducing their potential for increased revenue and changing the value chain.

Whether 4G, sometimes known as 'long-term evolution' (LTE), can help operators address these challenges and indeed how it might change the cellular industry is the focus of this chapter; other possible solutions such as femtocells are discussed in subsequent chapters.

2.2 Economics of data across cellular networks

Cellular operators around the world are now offering low-cost data tariffs. The exact value depends on the operator and changes over time but at the time of writing in 2010 is often in the region of between 3 and 15 GBytes/month for about $20–$25 with additional per-MByte charges if this limit is exceeded. In some cases, the tariffs were offered to fill up capacity that would otherwise be unused on a particular network. Once a network has been built, the marginal cost of using the capacity is almost zero and so strategies that 'fill up' the network capacity make sense. However, if this is the case, then the tariffs might not be high enough to fund additional investment. This section sets out some simple estimates of network cost to show that there may indeed be a problem.

Determining network capacity accurately is extremely difficult. The capacity per cell depends on factors such as the network engineering, the profile of the services offered and the speed at which the handsets move. The network capacity depends on the number of cells, the extent to which they are capacity- or coverage-limited and whether they are correctly located for the demand pattern. Operators can, to some degree, reduce demand by decreasing the data rates provided. Hence, any estimate of the capacity of networks will necessarily be highly approximate.

Cellular networks have enhanced their capacity with upgraded packet-based technology such as the high-speed downlink packet access (HSDPA) enhancements applied to 3G cellular networks. These broadly enable higher data rates for those with good coverage while excluding those with poor coverage. The complexity of these systems makes it very difficult to derive their cell capacity analytically. Instead, Qualcomm and others have modelled and measured typical scenarios and concluded that data rates in the region of 1.2–1.5 Mbits/s per sector per carrier can be supported using 3G high-speed data solutions on average.[1]

Let us take the UK for analysis as an example of a country with highly competitive cellular provision. Assuming about 10,000 cells covering the UK, each with three sectors, the total data capacity per operator is in the region of 40,000 Mbits/s or 40 GBits/s across the entire network – less than can be delivered across a single optical fibre. Operators have sufficient spectrum for two or three carriers, so the maximum capacity could be about 120 GBits/s.

[1] This may seem low given that HSDPA offers data rates of 3 Mbits/s or higher. However, these rates are only available in a very small number of locations and are rarely achieved. This also demonstrates that if a user does get 1–2 Mbits/s download speeds they are likely using most of the capacity of that cell.

However, many of their cells will be deployed to provide coverage rather than capacity and so will not be fully utilised. So, simplistically, something in the region of 30–40 Gbits/s might be more reasonable.

This capacity needs to be divided between voice and data users. If we assumed that about 50 million UK subscribers were evenly distributed across the five UK networks then there would be 10 million subscribers per network. Ofcom research suggests about 100 billion minutes of mobile calls per year – 20 billion per network or 55 million per network per day. Assuming an even distribution over the 14 hours from 8am to 10pm, this is approximately 4 million per hour. A voice-call minute uses about 720 kbits, so 4 million minutes equates to almost 3 Gbits/s. This clearly demonstrates why a well-developed 3G network has ample spare capacity for data – voice uses only about 10% of its maximum available capacity (and, of course, not all voice subscribers use 3G as opposed to 2G, so the estimates here are on the high side). So, after allowing for voice traffic, there might still be about 30 Gbits/s of available network capacity (SMS traffic levels are negligible).

Now consider data usage. If a subscriber with a large 15-Gbyte (which is 120 Gbits) bundle were to use all their data, this would equate to 4 Gbits/day, or, assuming the same 14-hour distribution of data, 280 Mbits/hour, which equals 80 kbits/s. Of course, most subscribers will use less than their allowance, so this is an upper margin. For the moment let us assume they use half their allowance at 40 kbits/s.[2] This means a network with capacity 30 Gbits/s can support 750,000 subscribers. The five UK networks can support a total of just under 4 million users.

On the basis of feedback received from mobile data users, this appears about right. A number of users of 3G data cards have commented on how the speeds they are getting have fallen dramatically between 2008 and 2009 from in excess of 1 Mbit/s to only tens of kbits/s, suggesting that networks are running into congestion problems. Taking into account the fact that traffic distribution is invariably uneven across the network by assuming a 3:1 peak-to-average utilisation, then, in some areas the network would already have reached capacity. Hence, the anecdotal comments on congestion would fit the approximate calculations set out above.

So the number of mobile data subscribers might be able to at most double from current levels before congestion problems start to become sufficiently serious to hinder further uptake.

[2] As a point of reference, a subscriber using half their 15-Gbyte allowance would be able to download about 30 hours of PC-quality video per month or about 10 hours of TV-quality standard-definition video.

Once the operators reach congestion, they can expand capacity by building more cells or by deploying additional carriers. The latter can happen only if they acquire more spectrum, perhaps at 2.6 GHz. However, there would be a time delay of perhaps 3–4 years to acquire the spectrum, for the manufacturers to develop 2.6 GHz base stations and subscriber devices, and for the base stations to be deployed into the network and the subscriber devices to become widespread amongst consumers. There would be costs associated with spectrum acquisition, the addition of new radios to base stations, device subsidy and a possible need for more cells, since coverage at the new frequencies (likely to be higher than current frequencies) will generally be inferior to that of existing 3G cells. In any case, additional spectrum will not arrive soon enough to resolve the current problem.

Building additional cells can occur somewhat more quickly, although the approvals process for new sites can take up to a year. However, it is costly. New sites in congested areas might cost $80,000 to $150,000 each and then have annual running costs at 10%–20% of these levels for backhaul and rental. A fully populated three-sector, two-carrier site would add about 8 Mbits/s, enough to support 100 data users under the most optimistic assumptions, including building the cell to precisely cover the hotspot and assuming an even traffic loading. This puts the cost of network expansion at about $1,500 capital expenditure (capex) and $150 annual operational expenditure (opex). Amortising the capital over 10 years, this is about $320 per subscriber per year or $25 per subscriber per month. Adding in costs of subscriber acquisition, datacard subsidy and other overheads suggests a need for roughly $50 per subscriber per month as a minimum fee. If datacard users start to churn at the same rate as voice users, monthly fees may need to be even higher. Prior to the rapid growth in mobile data, it was fee levels in this region that were seen by many as the main stumbling block to penetration.

There is one further problem with data as the numbers above hint at. A network with 10 million voice users will need 3 Gbits/s of capacity but one with only about 750,000 data users will need 10 times that. Data is currently being sold at around 100–200 times less per bit than voice, despite the fact that the cost of transferring voice is the same as that for data. This leads to potential arbitrage opportunities. In particular, a datacard user will increasingly be tempted to use VoIP to make voice calls 'free' over their datacard. As smartphones become more widespread, the possibility of doing this will grow despite the best efforts of the operators to disable such functionality in devices they subsidise. If this causes voice revenue to fall while datacard subscriptions grow, the operators will see falling revenue coupled to increased network build costs. Over time these differentials in the rates for voice calls and data will have to be reduced. Voice call rates can only fall slowly if operators are not to see income tumble, suggesting that data rates will need to rise to prevent such arbitrage.

These calculations shed some additional light on the use of mobile networks to provide broadband access to the home. The data-usage levels assumed here would be low for likely future broadband use – allowing less than 30 min/day of video viewing. We might expect a factor of between a doubling and a 10-fold increase in likely usage for home broadband access compared with mobile broadband. With costs per subscriber of at least $50/month calculated above for network enhancements to enable mobile broadband, home broadband access would require costs of $100/month or substantially more. Hence, unless there is spare network capacity in the cell where the potential subscriber resides, the costs of broadband home provision via mobile would appear to be substantially more than the market might be expected to bear and uncompetitive compared with DSL or cable. Even if there were spare capacity, only about 10 subscribers per cell would be sufficient to use this capacity.

This position would change in the longer term when substantially more spectrum became available, allowing more carriers to be deployed. However, by this time data requirements might have also grown, particularly if high-definition (HD) video became popular.

In summary, the current position on mobile data appears unsustainable. Accurate calculations are near-impossible and there is the potential for substantial error, but it would appear that operators will soon be suffering serious congestion unless they take action. However, the cost of such action suggests that datacard subscription costs will need to rise substantially. In the longer term, as more spectrum comes onto the market, there is the potential for a three- to four-fold increase in network capacity, but it will be costly for operators to buy the spectrum and deploy the additional carriers.

2.3 Features of cellular networks

Before entering into a discussion as to the merits of the next generation of cellular networks, sometimes known as 4G and sometimes as long-term evolution (LTE), it is worth recalling some of the key facts about cellular networks. Networks are constructed to deliver coverage and capacity and to enable applications. Coverage is achieved by deploying base stations around the country. Capacity is a function of the amount of spectrum, the number of base stations and the efficiency of the technology (in terms of bits/Hz per cell). Historically, the key gains in capacity have come from the deployment of many more cells, with only small gains being achieved from additional spectrum or enhanced efficiency. As well as the basic network capacity, a network must also meet a range of other requirements. The best publicised of these is the data rate available, but delay and latency can also be critical in some applications.

New generations of cellular systems have been introduced for a range of reasons. The move from first-generation (analogue) cellular to the second generation (GSM) enabled a number of problems to be resolved, including security, lack of harmonisation across Europe and an inability to carry data. Capacity gains were also achieved, although this was only one of the drivers for change.

The move from second- to third-generation (3G) systems was originally seen as a mechanism to increase the peak data rate from about 200 kbits/s (achieved with the EDGE enhancement to GSM) to 2 Mbits/s (achieved with the HSPA enhancements to 3G). As with 2G, capacity increases were also expected, although these were not seen as the main driver for change. After many years, 3G did deliver the higher data speeds promised, although the benefits of this for the operators and users are mixed, as discussed further below.

For many years the higher data rates of 3G remained unused. Even the additional capacity it provided was little used, with operators often running their networks well below capacity and making little use of all the 3G frequencies that they had been awarded. That situation changed with the successful introduction of 3G datacards. Datacards had long been available, but their usage was low because the fees were relatively expensive – often in the region of $100 per month for a restricted amount of data. Such high fees reflected the operator's cost of providing the service, but tended to result in users preferring WiFi or other methods to gain data connectivity. That position changed in 2008 when some operators started to offer much discounted datacard usage, in the region of $20 a month. This approximately covered the marginal cost of using their spare capacity, although it provided insufficient revenue to cover the investment cost of building additional capacity. At this price, datacards became successful, with subscription numbers growing rapidly. The price of success was increasing congestion as users rapidly filled up available capacity, but operators were gaining insufficient revenue to justify investing in capacity enhancements.

It is against this backdrop that LTE is being trialled.

2.4 Key features of 4G

Compared with 3G, 4G technologies such as LTE feature a modified air interface that uses orthogonal frequency-division multiplexing (OFDM) rather than code-division multiple access (CDMA) and an all-IP core network instead of the mixed circuit- and packet-switched environment used by 3G. The aim of LTE is to deliver data rates of about 10 Mbits/s to end users and a lower cost per call to the operator as a result of the consolidated core network and greater capacity per carrier. This section considers whether it is likely to meet these goals.

One of the difficulties with wireless transmission is inter-symbol interference (ISI). A radio signal will follow multiple paths from transmitter to receiver, with perhaps a direct path and then a number of others formed by reflections from objects such as buildings and mountains. These other paths are longer than the direct path and so the signal takes longer to arrive at the receiver. The effect is the reception of multiple copies of the transmitted signal, each delayed and attenuated by differing amounts. If the data rate is relatively low then the duration of each transmitted piece of data (or 'symbol') is sufficiently long that all the reflected signals arrive before the next piece of data. However, as the data rate increases, the symbols become shorter and some of the reflected paths transmitted during one symbol arrive when subsequent symbols are being received. This interference from one symbol to subsequent symbols – known as inter-symbol interference (ISI) – can be highly problematic if not corrected. As data rates increase, symbols become shorter and ISI becomes an increasingly serious problem.

ISI can be corrected using an equaliser. The transmitter periodically sends a known sequence (often called a 'sounding sequence'), which allows the receiver to deduce the delay and attenuation associated with each reflection. It can then remove the effects of the reflections from the subsequently received signals. Equalisers are relatively complex devices and their complexity tends to rise exponentially according to the number of symbols over which the ISI is spread. Hence, there comes a point, even with an equaliser, where the ISI cannot be removed and the errors on the channel rise rapidly. Exactly where this point lies depends on the topography (which determines the reflections) and the state of the art of silicon implementation, but it is typically somewhere in the region of 1–10 Mbits/s. With 3G data rates towards the bottom end of this range and projected 4G data rates at the top end, or above, an alternative approach to using an equaliser was needed for 4G.

This is the key reason for the selection of OFDM. The effect of dividing the data stream into a number of subchannels is to have a much lower data rate (and hence longer symbol duration) on each subchannel, removing the need for complex equalisation. However, as always in wireless, gains in one area come at the expense of losses in another. OFDM requires a guard interval at the start of each symbol to allow time for reflections from the previous symbol to decay away, which reduces the spectral efficiency. It is also prone to fading on each of the subchannels when the multiple paths add up in such a manner as to cancel each other out. This fading cannot readily be removed and hence more powerful error correction is needed to overcome the errors that will occur.[3] Once appropriate measures are in place to combat these issues then 4G, using OFDM, can be extended to much higher data rates than 3G, predominantly through the use of broader-bandwidth channels (up to 20 MHz, compared with 5 MHz for 3G).

As a result, OFDM is not materially more spectrally efficient than 3G, so it cannot be expected to substantially increase cellular network capacity (other than by being deployed in additional spectrum, but if other technologies were deployed here they would also bring similar gains). It is not materially more cost-effective and indeed the costs of upgrading from, say, 3G to 4G are likely to outweigh any of the cost savings that might be achieved from broader-bandwidth carriers. LTE may bring some benefits in terms of global harmonisation – unlike for previous generations, there appears to be only a single standard for 4G. This will benefit operators who currently have diverse technologies across their different national networks.

LTE is also completing the slow process of moving the core of wireless networks from circuit switching to packet (or IP) switching. This started in 2G systems when the circuit switch was augmented with a packet-based system to handle data traffic generated by GPRS and then EDGE data facilities. This approach was continued in 3G, with packet switching used for data while circuit switching tended to continue to be used for voice traffic. Within LTE the aim is for all traffic to use a packet-switched core.

For the operator a move to an all-packet core should result in lower costs because only a single core switch type will be needed. However, operators will probably still run 2G and 3G networks in parallel for some time and the costs of the core network are typically only a small part of the overall network costs, so any savings will be relatively small. Core IP networks also promise the simpler introduction of new services, since these can now be implemented on servers sitting outside the core network rather than requiring changes to the software of the core network itself. Equally, this might allow third parties to implement services outside the operator's network altogether – indeed, recent experience suggests that this is more likely to be the outcome.

Having said this, although voice can be handled as a packet data service using voice over IP (VoIP) or similar protocols, many mobile operators have found this difficult to introduce. VoIP calls can take a long time to establish and can have a lower voice quality than circuit-switched calls in the event that packets get

[3] However, more powerful error correction requires a greater degree of redundancy to be added to the signal, reducing the spectrum efficiency that can be achieved.

delayed. Although there have been many initiatives to move voice to packet transmission, such as the intelligent multimedia service (IMS), they have not been successful due to the difficulties of making them work and the relatively small perceived benefits. Hence, within 3G systems voice calls have tended to be handled over the circuit-switched core while data has been handled by the packet-switching elements. That possibility does not exist for 4G, which gets some of its gains from optimising the radio system around an all-IP service. Hence, for 4G operators will either have to use IMS-type services to implement voice or will have to carry voice over different networks (e.g. 2G or 3G).

At the time of writing in 2010, industry was split over whether it was better simply not to support voice within 4G and rely on 2G/3G fallback or whether the effort should be made to produce appropriate standards for 4G. This raises the possibility that in the future there might be separate networks optimised for voice and data, which might seem rather a retrograde step in a world of increasing convergence. Such a step would also imply that operators would not be able to fully decommission older networks, which would have implications both for their cost base and also for their ability to achieve gains in spectrum efficiency from the newer technologies. It would also limit the ability of new operators to enter the market with a 4G network unless they could secure favourable agreements with existing operators to enable subscriber roaming for voice.

For the end user the move to a packet core might have few immediate implications. The design of LTE allows for more efficient and faster handling of IP traffic than 3G, resulting in lower delays and thus enabling applications such as gaming that were not viable on 3G networks to work over wireless. The greater flexibility might also see the introduction of more services.

2.5 MIMO

One means whereby 4G might be able to achieve an enhanced spectrum efficiency over that of 3G is the use of multiple-input multiple-output (MIMO) antennas. MIMO is a term used to describe an approach whereby both the transmitter and the receiver have more than one antenna. A different signal is transmitted from each antenna at, say, the base station, but all transmissions are at the same time and same frequency. Each antenna at the receiver will receive a signal that is the combination of all the transmissions from the base station modified by the parameters of the radio channel through which each transmission passes. The path from each transmitting antenna to each receiving antenna will be slightly different, involving different angles of reflection and path lengths, and as a result each receiving antenna will see a slightly different combination of the signals from the transmitting antennas.

If the characteristics (phase shift, delay and amplitude) of each radio channel from each transmit antenna to each receive antenna are accurately known then, mathematically, it is relatively simple to apply a 'matrix-inversion' operation and deduce what data was transmitted from each antenna. In principle, if the transmitter had n antennas and a different signal were transmitted from each, then, as long as the receiver had at least n antennas as well, the capacity of the radio channel would be increased by a factor of n.

In order for MIMO to work, there must be a reasonable degree of difference among the transmission paths – if they are all identical, then each of the signals received at the subscriber-unit antennas will also be identical and it will not be possible to invert the matrix (in mathematical terms, the matrix will not have an inverse) and hence extract the different transmitted signals. In situations where there is a direct line of sight between the transmitter and the receiver, the differences may be relatively small and hence MIMO would be difficult to operate. In most mobile environments there is rarely a line of sight to the base station and multiple reflections take place, hence mobile operation would appear well suited to MIMO. However, it is also necessary that each of the transmission paths is known with high accuracy – otherwise the solution to the matrix equation will be inaccurate and it will not be possible to recover the individually transmitted signals. In a typical mobile environment, the transmission channels are constantly changing their parameters as the mobile moves, or vehicles or people move in the vicinity. Finding out what these parameters are requires 'sounding' the channel – sending a known sequence from the transmitter and using knowledge of this sequence at the receiver to deduce what the parameters of the channel must be. Sending this sequence is an overhead that reduces the capacity of the system to send the user information. If the channel changes rapidly, then it must be sounded frequently, eventually reaching a point where any gains from multiple antennas are negated through the overhead of sounding the channel.

MIMO systems are also complex and the difficulty of implementation increases rapidly if there is ISI, requiring channel equalisation as well as matrix inversion. OFDM is better suited to MIMO than most modulation schemes since it effectively removes the ISI, as discussed above.

Another issue that would make MIMO difficult to operate is interference. If the receiver also sees signals from other base stations, this will add noise or uncertainty into the process of separating the wanted signals, potentially to the point that the matrix inversion cannot be performed correctly and signal recovery becomes impossible. However, cellular systems are often designed with some inter-cell interference as a result of maximising the capacity by reducing the reuse between different cells. Systems could be designed with less interference, but this would reduce overall capacity, negating some of the benefits of MIMO.

MIMO adds cost to the network. Each antenna typically needs its own RF chain of components, which are some of the more expensive elements in the handset and base station and for which costs do not typically fall as fast as those for silicon. On the base station more space is required on the mast to mount multiple antennas, which is both costly and unsightly. On the mobile device a means must be found to integrate multiple antennas into a small form factor, remembering that there are already multiple antennas for all the different radio systems and that interference and coupling between them must be avoided. For the handset designer MIMO is far from a simple proposition, although for the laptop designer the constraints are somewhat less, with more space and fewer radio systems.

While MIMO appears to have worked in a number of tests and trials, its performance in real-world deployments is still unclear. In channels that change quickly and with interference from other cells, the performance gains from MIMO are likely to be substantially below the theoretical levels, while the increase in costs may prove higher than expected. Much of the promise of LTE is based on the higher data rates and capacities promised by MIMO, but it is still far from clear whether these will actually be delivered.

2.6 Spectrum allocation and layered networks

In order to provide a high data rate, 4G systems must operate in a broad-bandwidth channel. Much depends on the propagation conditions and the proximity to the transmitter, but, as a rule of thumb, the channel bandwidth needs to be at least as large as the data rate. So, for LTE to deliver 20 Mbits/s, it requires 20-MHz-wide channels. The problem is that such broad spectrum allocations are often difficult to come by. Those developing LTE recognised this and built into the system the flexibility to operate with a range of channels with bandwidths as low as 1 MHz, but, as the channel bandwidth reduces, so does the peak data rate that can be achieved.

Those deploying LTE systems may well be faced with a trade-off. At lower frequencies, where the propagation is better, it will typically be harder to find large bandwidths and hence the maximum data rates that can be provided will fall. Only at higher frequencies, where cells are smaller, will sufficient bandwidth be available to achieve the target data rates. Adding to the complexity, many operators will already have 2G and 3G systems deployed, offering a range of data rates and coverage.

The assumption is that layered networks will emerge, with lower-frequency systems used to provide good coverage and higher-frequency systems providing additional capacity and higher data rates in areas such as cities and key transport hubs. The actual balance will probably vary from operator to operator depending on their circumstances, spectrum allocation and views as to the costs and benefits of offering higher data rates. In general it might be expected that there will be 2G coverage in rural areas, providing up to about 100 kbits/s, 3G coverage in suburban areas, with data rates up to 1 Mbits/s, LTE coverage in urban areas, offering up to 10 Mbits/s, and WiFi in hotspots, offering perhaps even higher data rates, but more critically offloading the traffic from the cellular network.

2.7 The role of LTE

When design work started on LTE the assumption, as with 3G, was that the key reason for the new standard was to enable higher data-transfer speeds. The 3G standard was assumed to deliver data rates in the region 1–10 Mbits/s and LTE was seen as a way to extend this to the range 10–100 Mbits/s.

While higher data rates are always a good thing, many operators now see alternative reasons for the introduction of LTE, including the following.

- Higher capacity. Predominantly this comes from deploying LTE in new spectrum rather than replacing existing networks, helped by the expectation that 4G will be more spectrally efficient than 3G.
- Single technology. For some global operators the cellular networks they own around the world use a variety of 3G standards. LTE is seen as a way to harmonise on a single technology, bringing potential economies of scale and facilitating roaming.
- Better handling of data. LTE has been designed to have lower delays for data transmission, making it better suited to a number of data applications.

However, LTE comes at a difficult time for cellular operators. Cellular revenues are generally falling slightly, especially in developed countries, while at the same time data volumes are growing quickly. The resulting network congestion requires investment in additional capacity, but falling revenues suggest that further investment may be difficult to justify. Some operators are now questioning whether there is sufficient funding to enable competitive LTE networks to be deployed in a country or whether instead a single shared network to which all operators have open access would be a better model. Lack of sufficient spectrum for multiple LTE networks using 20-MHz-wide carriers in many countries is another factor making operators tend towards shared networks. Regulators tend to be concerned that shared networks may result in reduced competition, but, faced with an industry where consolidation is already occurring, may accept shared networks on the basis that this is better than no networks at all. Indeed, most regulators also regulate fixed access where there is often only a single network to which access is shared by multiple ISPs and virtual telecoms providers.

So, broadly, the role of LTE may be more about bringing more network capacity on stream in a relatively low-cost manner through sharing networks than it is about higher data rates or new features. In any case, the experience of 3G, which was first introduced in 2000 and some 10 years later is still with fewer subscribers than 2G networks, tells us that the move to LTE is unlikely to be rapid.

Reference

[1] http://www.ofcom.org.uk/research/radiocomms/reports/economic_spectrum_use/.

3

Femtocells

3.1 Rationale – home coverage and capacity

The capacity of any network is determined by the spectrum available, the efficiency of the technology and the number of cells. Historically, the vast majority of the gains in the capacity of cellular networks came from ever decreasing cell sizes. The logical conclusion of this is to implement small cells in each building – indeed, in some cases, in each room. If this were done, many have suggested that there would no longer be a capacity problem since many hundreds of Mbits/s would be available in each room, with only a few people in the room to share it between. For many years the cellular community has been working towards very small cells, partly to realise this vision.

Another reason for small cells is to enhance coverage within buildings. For most, cellular coverage indoors is adequate, but there are some situations where homes are on the edge of coverage or have particularly high building-penetration losses. A small cell in the home would provide high-quality coverage.

A final reason is associated with current business models. Cellular operators in many countries are different entities from the fixed-network operators. There is often competition between the two. When cellular users return home they may switch to the home fixed connection, perhaps because it is less expensive or the quality is better. Providing a home cell with associated modified billing could result in any call revenues from home being received by the mobile operator. This driver, however, is sometimes at odds with what might be in the best interests of the consumer and may change if the business models, ownership or structure of the mobile or fixed operators changed.

Coverage and capacity within the home could be provided by a small cell using a range of different technologies. At present WiFi and 3G cellular are two of the key contenders, but Bluetooth, LTE, cognitive solutions or others could be used. Given the large amount of capacity potentially available, it makes little difference from a functional viewpoint which technology is selected. The advantages of 3G are that it is

- already available within millions of handsets,
- can be subsidised by the operator and
- fits the current business models well.

WiFi has the following advantages.

- It is already deployed in many homes.
- It is independent of any operator, so it can be deployed by the home owner as they wish.
- Data rates are higher and more bandwidth is available than for cellular.

In practice, it seems likely that both technologies will be deployed in many homes, perhaps integrated into the same box.

In this chapter we concentrate on the more radical way of providing coverage in the home – that of femtocells. Towards the end of the chapter we will assess the likelihood of their deployment.

Before covering some of the technical and deployment issues, it is important to discuss the business case for a femtocell. The benefits of femtocell deployment fall both to the operator, who improves the coverage and capacity of their network, and to the user, who gains better coverage within their home. The costs are that of the femtocell itself (projected to fall towards $100 but currently somewhat higher than this), the backhaul (which is generally the broadband connection from the home paid for by the home owner) and a small on-going maintenance cost.

The current view is that the home owner will generally not be prepared to pay for the femtocell unless they have a particular need for enhanced cellular coverage within their home. This may change as applications develop or as operators provide lower tariffs for calls within the home (although in this case, arguably, the operator may be effectively paying for the femtocell through lower call revenue). Hence, the operator may need to partly or fully subsidise the femtocell if they wish to see widespread deployment. Operators may be prepared to do this in order to increase customer loyalty (reduce churn) and to enhance the coverage and capacity of their network at relatively low cost.

3.2 Interference

The basic concept of a femtocell is to place a small cellular base station inside the home (or other building) and for the inhabitants of that building (and perhaps visitors) to hand off onto that cell when within the building. For the cellular user the experience should broadly be the same as any other handoff except that additional services might be offered, such as lower call costs or particular applications triggered by detecting that the user is in the home.

For the femtocell to work with existing handsets it must operate in the existing 3G spectrum. In most countries that has already been fully licensed to operators and hence femtocells must operate in a given operator's spectrum. Because this might cause interference, it can happen only under the control of the operator. Hence, femtocells cannot be deployed on an unlicensed basis, unlike WiFi, but are typically bought from an operator and used as part of a mobile subscription with that operator. The femtocell is then typically installed by the user by plugging it into the home broadband connection, where it is able to contact the cellular infrastructure and obtain configuration information.

A key question for the operator (and to some degree the home owner) is whether the femtocell should be open to others not resident within the home or closed such that only a pre-defined set of users can access it. There are many dimensions to this question, including interference, congestion relief and user preference.

- *Interference.* While the signal from the femtocell is designed to cover the building, generally some will leak outside. There may be other mobile users from the same operator in the vicinity of the building who will receive the signal. If the femtocell is open they may choose to hand over to the femtocell (although if they are travelling fast this may be problematic). However, if the femtocell is closed they will be unable to use it but may suffer interference, especially if they are close to the edge of their serving cell. Modelling suggests that in this case the interference from the femtocell might result in a coverage 'hole' with a radius of about 30 m around the building. Most operators have more than one 3G channel and so a possibility may be to hand over to a different channel, assuming that one is available. Adjacent-channel interference issues are much less problematic, with coverage holes being only of size about 7 m at the cell edge, barely larger than most homes. Hence, while openness will solve the interference issue, it can be mostly mitigated by using more than one carrier across the network.

- *Congestion relief.* From the operator's point of view, one of the key advantages of femtocells is to carry data traffic that otherwise might result in congestion on the macrocellular network. If the femtocells are open then they can also capture the data traffic of other users in the vicinity. For many homes this might not be relevant – perhaps few users pass close to the home or do so at sufficiently low speed that handing over to the femtocell is viable. However, there may be a limited number of locations where enabling access from other mobiles can offload significant data traffic from the network.
- *User preference.* Users might not wish for femtocells they have installed in their home to be open to strangers. They may perceive that the stranger (or the operator) is effectively making free use of their backhaul connection or be concerned about issues of privacy or security. If there is some reciprocal gain for them, such as the ability to access the femtocells of others free of charge, then they might be more inclined to agree. It is still unclear what the views of users will be in this respect.

Femtocells could go a stage further than the openness described above. As well as being open to others on the same network, they could conceivably support mobiles from a range of networks. This might be attractive to the home owner, since all the mobiles in the home would not need to be on the same network, or it might be attractive to the owner of a public building who would need to install just a single device to provide service to all those entering the building. For this to work, either the femtocell would have to transmit multiple carriers, one corresponding to each mobile network, or roaming agreements allowing mobiles from one operator to roam onto a network operated by another would be needed.

While both are technically possible, neither fits well with the business case for femtocells where the operator subsidises the femtocell. It does not make sense for one operator to subsidise a femtocell when the benefits of its deployment might accrue to its competitors. Equally, roaming between the networks of different operators in the same country has typically not been implemented since the operators do not perceive it to be in their commercial interests and regulators generally do not perceive it to be appropriate to mandate it.[1]

[1] Regulators are generally wary of mandating roaming because it could reduce competition and remove the incentive for an operator to increase the coverage of their network since they could not claim greater coverage than their competitors in doing so because their competitors would simply roam onto any new cell. Perversely, the result of mandating roaming might be to result in worse coverage in the longer term as a result of the lowered incentive to enhance coverage.

Therefore, this greater degree of openness looks unlikely at this point. This might change in the future if there were network consolidation or other changes in the business structures, but until it does it would not be possible to use a particular femtocell unless the user subscribed to that particular operator. This may make femtocells unattractive to those households where there are multiple mobile phones.

3.3 Handoff and registration

Femtocells are not easy to integrate into 3G networks, primarily because the design of the way devices are registered and hand over to other cells was made on the assumption of networks where all the cells were under the direct control of the operator. In order to find a cell to hand into, the mobile needs to know what to look for – this is often provided to the phone as information in the 'neighbour list'; and in order to make the handover substantial signalling traffic needs to flow between various elements in the network. Neighbour cell lists are typically populated by the operator as they build and design their network, but it is difficult to incorporate femtocells into this process because the operator might not know exactly where they are. Further, there may be too many of them within a larger cell for the mobile to successfully look for them all (and in any case, the mobile may only be able to register onto its own femtocell, not onto those provided by other homes).

For these reasons, many femtocell implementations do not support handover in at least one direction, typically to the femtocell from the macrocell, and sometimes also in the other direction. However, femtocell vendors claim to have solved most of the other integration problems with automated self-planning devices that remove from the operator the complexity of potentially having to manage a network with millions of devices.

3.4 Commercial choices

For femtocells to be successful, there must first be a decision by the operator to offer them and then a decision by users to adopt them within their homes. Because of the need for femtocells to operate on licensed spectrum, users cannot adopt femtocells without the prior approval of the operator.

The decision for the operator is a complex one. Deployment of femtocells might reduce churn and offload some of the data traffic from their network, both of which will bring commercial benefits. However, there will be costs associated with any subsidy of the femtocell, additional network elements and upgraded customer service, and there may be some interference from femtocells or other network problems. It is difficult for operators to build an evidence-based business case when factors like churn reduction are not known, femtocell costs in the longer run are unclear and many other factors cannot be quantified. A further complexity for the operator is whether 3G femtocells should be deployed or whether they should await the arrival of 4G in order to future-proof their deployment as much as possible.

With much uncertainty, the operators have tended to opt for trials and in a few cases for low-key initial deployments with limited subsidies. This will allow them to gather experience with femtocells prior to a large-scale launch. The success of these original trials will determine whether operators decide to promote femtocells. It may well be an 'all or nothing' result in that, if one operator decides to adopt femtocells, then its competitors might also follow in order that they remain competitive. Conversely, if the operator leading in femtocell trials decides not to go ahead, then its competitors may decide they no longer need to trial or consider femtocells. This decision will hang in the balance in the period 2010–2012 and it may be only towards the end of this period that it will become clear whether femtocells will be successful.

Even if operators promote femtocells, users have to make a choice to adopt them. Their attractiveness to the end user will be better cellular coverage in the home and the ability to use the mobile in the home on a reduced tariff. The downside may be some additional cost and the need for an extra wireless box in their home, which requires power and connection to the broadband network. Each user will have differing requirements and perceptions as to the value of femtocells.

One key factor will be the price at which operators decide to sell femtocells. At one extreme they might mark up the price at which they acquire femtocells from the manufacturers to cover all their costs including increased customer care. At the other, they might heavily subsidise them in the same way as many handsets are subsidised in order to attract and retain customers. Femtocells might be linked to particular subscription offers designed to encourage an entire family to move to one operator, with multi-year contracts intended to reduce churn.

Another factor will be the way that communications within the home develop. An alternative to femtocells is to use the home WiFi network in order to make voice and data calls within the home. This is becoming increasingly possible as phones progressively add WiFi to their capabilities and as methods for making VoIP calls, such as Skype, become increasingly prevalent. At present, there are few mainstream companies offering this capability, but this may change over the period during which femtocells are introduced. Hybrid solutions may emerge whereby a mobile phone uses the home WiFi network for data transfer but remains on the 3G macrocellular network for voice calls.

The group structure of the cellular operator may also make a difference. If a cellular operator is linked to a fixed operator, perhaps through common group ownership, then they will be less concerned about calls within the home being carried across the fixed network and hence the incentive for femtocells may be less. The converse will be true for mobile operators with no link to fixed operators.

All of this leads to a conclusion that the future of femtocells is too difficult to forecast at the moment. They could be a complete failure or highly successful, and the factors that could tip the balance between these two outcomes are very difficult to forecast and may depend on bold decisions taken by a few key individuals. It seems certain that most homes in the UK will have some form of wireless base station within the next decade, but whether this will be WiFi, femtocell or both is too hard to forecast.

3.5 Verdict

The future of wireless is small cells in most homes and buildings providing very high capacity. At present, most of these small cells are based on WiFi, providing data coverage. The question remains as to whether they might be replaced or augmented by femtocells. We have argued that this is a very difficult call, depending on a wide variety of unknown factors that could tip it one way or the other.

In such a situation any verdict must be treated with much caution. On balance, we believe that the costs of femtocells will outweigh the benefits for most operators and that, with a home WiFi network already installed, users will not be keen on putting additional wireless networks into their home, especially one that remains under the control of the operator. Hence, we do not expect to see ubiquitous femtocell deployment. However, the jury is still out on this one.

4
Cells in the sky

4.1 The benefits of cells in the sky

Cells in the sky can broadly be divided into high-altitude platforms (HAPs) and satellites. HAPs are based on flying platforms such as aircraft or balloons, operating at altitudes of up to about 60,000 feet, while satellites use a wide range of orbits from low-Earth-orbit (LEO) systems such as Iridium at 300–800 km above the Earth to geo-stationary (GEO) satellites such as those used for TV broadcasting at 36,000 km above the Earth's surface. Systems of each type have their own particular characteristics, which will be discussed in the following sections.

A cell in the sky can provide excellent outdoor coverage. Owing to its elevated position, obstacles such as mountains or buildings tend not to get in the way, allowing line-of-sight propagation from many locations. Large cells can readily be provided, enabling coverage of both urban and rural areas. Because much of the propagation is line-of-sight, higher frequencies, such as those above 3 GHz, can be used. These are inappropriate for cellular communications because the decrease in reflection and refraction at higher frequencies prevents good coverage, but this is generally not a problem for cells in the sky. At higher frequencies spectrum is both less expensive and more plentiful. A good example of this is TV broadcasting. Terrestrial TV broadcasting uses frequencies in the UHF range – about 500–800 MHz. These frequencies are scarce and expensive but, despite their greater coverage, about 1,000 cells are required in order to cover a country the size of the UK with a limited number of channels. Conversely, satellite TV broadcasting covers the entire country with one satellite using frequencies at 12 GHz. With much more spectrum available in this band, orders of magnitude more channels can be broadcast. For broadcasting applications, satellite transmission appears much more appropriate than terrestrial.

Cells in the sky also have problems associated with them. One problem is the cost of getting the cell into the sky and, in the case of HAPs, keeping it there. Another is poor indoor coverage as a result of the fact that signals are travelling vertically downwards and would generally need to pass through the roof and any upper floors of the building before reaching the receiver, although in some cases enough signal power may pass obliquely through windows to enable communications. A third problem is backhaul, since the signals to be transmitted must be passed up to the cell in the sky and the received signals passed back down. Standard wireless backhaul solutions are typically not applicable to cells in the sky. A final problem can be capacity. Because transmitters in the sky create large cells on the ground, their capacity is relatively low compared with that of conventional cellular systems. This can be improved to some degree with multi-beam antennas, but there are practical limits to what can be achieved here.

To date, satellite systems have played only a minor role in communications networks (with the exception of broadcasting satellites). HAPs are rarely used, with no commercial deployments but possibly some military adoption.

4.2 Satellites

Satellites have been used for communications for some 50 years. Initially they provided a mechanism to enable communications over large distances, forming trans-Atlantic links. They then found a core role in providing communications where terrestrial infrastructure was not feasible, such as in providing communications to ships and aircraft flying over water. Indeed, this role of maritime communications gave its name to one of the major satellite-network operators – Inmarsat. This role continues and is growing (see Chapter 17) as passengers on ships and planes increasingly look to have Internet connectivity. As a result, many forecasts suggest that satellite capacity for maritime and aviation use will be insufficient by about 2012 unless new satellites are launched. Such satellites require a small dish at the receiver, which is oriented towards the satellite in order to gain a satisfactory signal. This allows portable units (often called very-small-aperture terminals or VSATs), but not ones that can be used without being unfolded and pointed in the right direction.

As mentioned before, satellites also play a major role in broadcast transmissions, for which their signals are typically received by a satellite dish mounted on the side of the house and oriented towards the satellite. Satellites also play many other roles, such as observing key parameters on the Earth, including temperature and sea height, providing weather-related information such as cloud cover and assisting radio astronomy, which we will not cover in this chapter.

While useful for communications to ships and aircraft, satellites have not played a role of any significance in mobile communications. This is because mobile devices cannot deploy large directional antennas towards the satellite, with the result that the signal they can receive is generally too weak. One solution to this is to move the satellites closer and this idea resulted in the launch in the 1990s of the LEO satellite systems, most notably Motorola's Iridium. With the satellite much closer to the Earth, smaller, although still rather ungainly, antennas could be used. Delays to communications were also reduced. However, about 70 LEO satellites are needed for global coverage, compared with about 3 GEO satellites, and LEO satellites are in unstable orbits requiring their relatively frequent replacement. The cost of launching and maintaining the system proved much higher than the revenue that could be derived and Iridium went out of business. As an aside, its assets were purchased by the US military amongst others and it has now found a role as a valuable military communications tool, especially in areas such as Afghanistan.

A more recent proposal for satellite-to-mobile communications is the so-called 'complementary ground component'. In this architecture, conventional base stations receive the signal from the satellite (using a dish oriented towards the satellite) and then re-radiate from a cellular-like antenna. Such an arrangement is not materially different from a cellular network where the backhaul from the cell sites is achieved via satellite rather than microwave or fibre. Satellite is generally less appropriate as a backhaul technique, so it is hard to see why such a concept should bring any material gain.

Satellites have also long been suggested as an alternative means to provide broadband connections to the home. In this role, directional antennas can be deployed on the sides of houses, so the problems associated with mobile communications with satellites are not relevant. Instead, the issue shifts to one of capacity. Broadband connections often provide data rates of 2 Mbits/s or more and the amount of data transferred is growing rapidly as users stream video content. Satellites have similar capacity constraints to cellular systems in that capacity is a function of the spectrum allocation, number of cells and technical efficiency. In the case of satellites, additional cells can be formed not by launching additional satellites but rather by deploying antennas on the satellite that create a number of 'spot beams'. It is typically possible to reuse the frequencies in each spot beam, or sometimes using a frequency-repetition pattern. More spot beams require larger antennas, which are problematic on satellites, especially at launch time, and also require more processing on board the satellite, which may cause power-consumption issues. Despite this, the number of beams has been steadily increasing, with recent satellites having some 800 'microcells' or spot beams formed by using antennas with some 2,000 elements. Apart from the problems of size and power consumption, keeping these spot beams stable on the Earth can be problematic. If they drift too much, then subscribers may move from one beam to another and either temporarily or permanently lose service. Nevertheless, as technology steadily improves, the number of beams can be expected to increase.

Even with 800 beams, satellite capacity is still relatively small. Such a satellite might cover most of Western Europe or the USA, where a cellular network would have some 200,000 cells (not counting femtocells and WiFi hotspots) and even cellular networks have insufficient capacity to provide broadband home connections to a large percentage of the population. For these reasons, satellites can form only a part of a broadband-coverage solution, providing service in rural areas where there are no other alternatives.

So, in summary, satellites will continue to play an important role in providing communications to ships and aircraft, in addition to a role for military use and a small role in providing broadband communications to the home. While satellites will steadily improve, no new developments that will materially change this are expected.

4.3 High-altitude platforms

HAPs are similar in many respects to satellites in terms of their advantages and disadvantages. Key differences include the need for an airborne platform, that they are closer to the ground than satellites and that they can be moved to different areas.

One serious problem with HAPs is keeping the base station in the sky for long periods of time (ideally indefinitely). Many options have been explored. Manned planes can be used, but are very expensive. Unmanned planes, often powered by solar cells, provide many advantages, but are still expensive and often can fly for only a day or two before they need additional power. Tethered balloons are less expensive, but the tethering cord can be a serious problem to those in aircraft and others. While there are many options, there are no fully satisfactory platforms that combine low cost, substantial payload and long flight time. Given that substantial research into aircraft and more specifically unmanned vehicles for military use has not resolved this problem to date, breakthroughs seem unlikely.

Beneficial, though, is that they are much closer to the ground than satellites. This allows higher signal strength than from satellites, making smaller antennas possible for mobiles. Indoor penetration typically still remains a problem. Another benefit of being closer to the ground is that the coverage area is smaller than that of satellites, allowing spot beams to be smaller and hence, if multiple HAPs are deployed, for the capacity to be higher than that for a satellite.

Capacity will remain lower than that of cellular systems because the spot beams will still generate cells much larger than those deployed by cellular operators in urban areas.

HAPs would seem to be useful in a number of special situations. One is military operations, where a rapidly deployable high-capacity relay in the sky can provide a wide range of battlefield communications. Communications in disaster zones is a similar application. Another, potentially, is the rapid provision of additional capacity to a cellular network in the case of an unusual and unexpected event (expected events are generally covered with transportable cells in trucks that can be taken to the venue). A HAP could be quickly flown to the area and provide additional coverage.

Outside these situations it is hard to see a key role for HAPs. Coverage is relatively expensive, and poor indoors, and capacity is lower than that of a well-developed cellular network.

4.4 Verdict

While satellites and HAPs are often discussed as mechanisms to provide mobile and fixed capacity, neither is a good substitute for cellular or fixed terrestrial networks. This is because the capacity they provide is relatively low and their coverage indoors, where much communication takes place, is poor. While antennas steadily improve, enabling more spot beams, there are no technological breakthroughs on the horizon.

Satellites are excellent in broadcast applications and are the only viable mechanism for providing connectivity to ships and aircraft. HAPs have important military applications. Beyond this it is unlikely that they will change the communications networks of the future.

5
Mesh networks

5.1 Introduction to mesh networks

Conventional wireless networks have a central transmitter, often termed a base station, transmitter mast or node. This controls the communications with devices within its range. For example, in a cellular system base stations provide coverage across an area and control the access from mobiles in the vicinity. The central transmitter is often elevated relative to the receivers – transmitters of cellular masts are typically 10–20 m above the ground while mobiles are mostly 1–2 m above ground.

A much discussed alternative is for there not to be a central transmitter. In the most extreme case devices transmit to other devices that relay their message onwards. If, for example, all communications occurred within a shopping mall, it might be quite possible for messages to pass from transmitter to intended recipient via re-transmissions (often termed 'hops') from one device to another across the mesh. Alternatively, the message might pass through a mesh in order to reach a point of interconnection with the fixed network (a 'sink node'). At this point the message would be routed through the fixed network to the recipient in a conventional manner, although the final 'drop' to the recipient might be via another wireless mesh network.

Mesh systems potentially bring a number of advantages.

- No need for infrastructure. Without any central transmitters, mesh networks do not require any infrastructure and hence are simpler, cheaper and faster to establish than conventional networks. They can also work where it is not possible to deploy a central infrastructure, perhaps in a war zone or during a civil emergency.

- Reduced range requirements. In a conventional system the devices must be able to communicate from the cell edge to the central transmitter, which might be many kilometres away. In a mesh system they need only communicate as far as the nearest other device, which might be only hundreds of metres. This could enable communications systems to be deployed where conventional systems would be uneconomic because of insufficient range, perhaps as a result of operating at high frequencies or with low power levels.
- Possible capacity enhancements. Some suggest that mesh systems can deliver greater capacity than conventional solutions, but, as we will see, this is true only under certain conditions.

However, mesh systems also have their problems.

- Uncertain message delivery. Because the route that a message might take is often uncertain at the point from which it is transmitted, it is possible that it might get delayed or lost. Few guarantees of service can be provided unless the mesh is static and engineered.
- The need for seeding. Mesh systems work only when there is a sufficient density of devices to forward messages. Initially this might not be the case and as a result it may be difficult to get the mesh started. One solution is to 'seed' the mesh with additional devices whose only function is to act as relays. However, there are cost implications to doing this.

Mesh systems have had only limited success to date. While mesh modes are available within WiFi, they are rarely used. In the late 1990s and early 2000s a few mesh companies were started, such as Radiant in the UK and MeshNetworks in the USA. These have had limited, if any success. BT trialled a version of Radiant's solution designed to provide broadband coverage in rural areas, but decided not to pursue it further. Most US companies were acquired by larger manufacturers such as Nokia and Motorola, and it appears that their products have become niche, at best. Plans to deploy WiFi systems across entire metropolitan areas using mesh backhaul have come to nothing, although this may be as much a reflection of the issues involved in providing such coverage with WiFi as it is of the mesh backhaul. It does appear that the military are using mesh systems on battlefields, but there is limited information available on this.

Perhaps the greatest success for mesh today has been some in-home systems. For example, companies such as Sonos offer an audio system with speakers in multiple rooms working off a single remote control that uses mesh networking around the home to link all the nodes together. Similar approaches may be used for other home systems such as monitoring and security solutions.

There is still much interest in using mesh to link sensor networks together (e.g. networks of temperature sensors distributed throughout a building). Sensors are discussed further in Chapter 13. A key problem with sensors is the need to use very low power levels in order to prolong battery life and mesh systems, with their potentially short link lengths, may be a mechanism with which to achieve this. Since sensor data is often not time critical and sensor networks can be planned and hence do not need seeding, many of the key disadvantages of mesh networks do not apply in this situation.

This chapter looks at the question of capacity and coverage in mesh networks as well as some of the complexities that need to be overcome.

5.2 How mesh systems work

Mesh systems are basically short-range radio devices with embedded algorithms that allow them to route information appropriately. The short-range wireless operation is quite simple but the routing is much more problematic. A mesh device may either originate information (if its owner has something to send) or receive information from other nearby devices. In either case, it needs to transmit this information onwards so that it gets closer to its final destination. In some cases mesh systems will contain both the sender and the recipient within the same mesh (for example, an emergency service network in a building), but in many cases the mesh will channel the information to a 'sink point' – a node that has connections to an external network, for example via a fixed broadband connection. Further routing can then be performed using the Internet or other standard routing mechanisms. In this section, we will use the term 'destination' to mean either the final destination or a sink node within the mesh.

Routing algorithms have been well developed within the Internet and are routinely used to route packets from computer to server and similar. They tend to work through a process of discovery, with nodes such as routers working out for themselves which other nodes they are connected to and which addresses those nodes are closer to than they themselves are. Packets are not always routed through the most efficient path, but the algorithms typically deliver routes that are close to optimal while not requiring manual intervention and adapting to changes in environment as nodes are added or removed, or suddenly become unavailable, perhaps due to a link failure. Such routing algorithms are beyond the scope of this book, but it should suffice here to say that they have been the subject of much research and ingenuity.

The key advantage that Internet routing has over mesh systems is that Internet nodes are static and they are introduced and removed on relatively long timescales. Hence, nodes have some time to build up and refine a picture of the other nodes with which they can communicate and the destinations that each node is closer to. This can often be done during 'quiet periods' so that each node has sufficient information that, when a packet of data arrives, it consults internal tables it has populated and forwards the data accordingly.

Some wireless mesh systems are relatively static – for example, those used for fixed wireless deployments. In these systems, Internet-style routing algorithms can be employed. Alternatively, central planning can be used, so that, as a new node is added (for example, when a subscriber orders a connection for their home), a central planning tool can determine which other nodes the new node should be able to communicate with and send round updated routing tables to all nearby nodes.

Such examples are relatively rare, though. Most practical mesh systems do not have the possibility of central planning and may have nodes that move around. Movement leads to a large increase in the complexity of the design of routing algorithms. Devices can try to maintain routing tables as they move, but typically this requires much overhead of radio resource transmission and, unless nodes are very frequently sending user information, such an approach is likely to be inefficient. Instead, routing must be done 'as needed' – when there is a data packet to transmit a node must ascertain what other nodes are in the vicinity and which of these can help forward the data to its destination.

Many approaches to solving this problem have been suggested and there is much literature on how mobile mesh routing can work. One simple approach is just to 'flood' the mesh so that a node sends on a packet of data to all the nodes that are within receiving distance. Each of these does the same and so on. Eventually the message reaches a destination node. If the mesh is relatively static, information on the route that 'worked' can be returned through the mesh and used to optimise the routing of subsequent messages, but this works less well when the mesh is constantly changing (when even returning such a message becomes challenging). This is simple but obviously very inefficient in its use of radio resources and battery power. Another approach is to hope that nodes physically move nearer the destination and so can 'carry' the message before forwarding. This is efficient in terms of radio resources but can lead to very large delays or potentially a complete failure of message transmission.

The difficulty and inefficiency of routing in mobile mesh systems is one of the key reasons why they are not able to deliver on their initial promise of increasing mobile capacity, as will be discussed in the next section.

5.3 Mesh and capacity

One of the claims made for mesh networks was that they achieved higher capacity than an equivalent cellular network – that is, more users could be supported within the same spectrum using mesh rather than cellular systems. This turns out to be a claim that needs considerable definition before it can be evaluated. There is no equivalent cellular system for a given mesh. We could envisage perhaps a single cell with a transmitter in the centre and a mesh with a sink point also in the centre. We would need to make assumptions about the cellular technology, the interference levels from surrounding cells and even the directionality of any antenna used on the cellular base station. A detailed description of how all this might be done is available in Methley [1].

The claim for higher capacity of mesh is based around the fact that the mesh network forms a set of short-range links, each of which is at relatively low power and hence causes little interference. This could be compared with a single cell that has been split into many smaller cells and it is well known that smaller cells generate more capacity – but this is true only in a conventional network where each cell has its own dedicated backhaul connection.

A device near the edge of the cell in the conventional system would set up a relatively long-distance link to the base station. This would generate interference with others and would probably have a relatively low capacity because higher-order modulation could not be used. In a mesh system, it would set up a short-distance link to the nearest device, which in turn would set up another link and so on until the information reached the sink node in the centre of the cell. Some relatively simple analysis shows that, if these hops were all of the same length and were all directly in line from the originator to the sink, then a mesh system would indeed have a higher capacity. However, the capacity gain falls if the links are of unequal length, since the longer link will generate more interference and not be able to use such a high number of modulation levels. The capacity gain also falls if the links are not directly aligned with the sink node, since additional hops are then required, further adding to the interference in the cell. In a real-world mesh neither equal-length hops nor alignment is very likely. If the routing algorithm is imperfect, then even more inefficiencies could result. Finally, cellular systems achieve a gain through the use of sectorised antennas on the cell site, which increase the link gain, giving an advantage to the cellular system that is not easily replicated in the mesh system.

The net result of this is that in most practical deployments it seems likely that mesh systems will be less efficient than cellular ones. The actual difference depends on many factors and assumptions, but is sufficient to suggest that mesh networks are not more efficient than cellular ones.

There is one way in which mesh can improve overall spectrum utilisation. Because mesh paths are shorter than those in conventional systems, mesh can operate in higher frequency bands. For example, cellular systems are generally not considered viable above 3 GHz, but mesh systems with short line-of-sight hops could readily work at 5 GHz and probably even higher than this. This opens up 'new' spectrum so that a mesh system could run in parallel with a cellular system, providing a collectively higher capacity than just the cellular solution. Of course, there are cost implications to this and it may be that simply increasing the density of cells in the cellular system is a cheaper way to achieve the same thing.

5.4 Mesh and coverage

Under certain situations mesh systems can usefully provide coverage where none existed previously. Broadly, these will be areas where providing cellular coverage is uneconomic but mesh might provide an alternative. There may also be a coverage role in situations where the cellular coverage has failed (perhaps because of a disaster) and mesh provides a backup. This section considers these areas.

Unlike cellular solutions, the coverage of mesh systems is not pre-determined (unless the mesh is static – a situation we will return to in the next section). A device has coverage only if it can see other nodes, and that coverage is useful only if those other nodes can form a path back to a sink node. Whether coverage exists at any point depends, then, on the number and distribution of nodes in the vicinity. Coverage may come and go as nodes move around and may vary over longer time periods as subscribers join or leave the mesh. Coverage is a particular problem at the start of a mesh deployment, as noted earlier, and seed nodes may be needed in order to address this. As a result, few operators would prefer a mesh solution as a means to provide coverage when conventional cellular coverage was viable. However, when cellular coverage is not available, mesh might provide a backup.

One possible role for mesh systems is to provide 'coverage extension'. In this approach a cellular base station attempts to cover the cell. However, there may be areas such as within buildings or on the edge of the cell where coverage is not available. A mobile in one of these areas might use a mesh-like solution to communicate with a mobile outside the building that had acceptable coverage. This mobile could then forward the call on to the cellular network. Such a solution might be relatively simple if just a single hop were involved.

While it appears attractive, there are problems with this approach. Technically there can be difficulties in identifying mobiles that can route messages and in maintaining the communications when they move. Commercially there may be issues of subscriber acceptance and battery life. These are discussed below.

Using existing mobile standards, it is very difficult for one mobile to identify another nearby. Mobiles transmit only when their user is active, or briefly if they move from one cell to another. If a mobile is relatively static and is not involved in a call then it need not transmit for some time. Either this would need to be changed or the mobile that was out of coverage would need to transmit an 'anyone out there' message. Both would require changes to the cellular standards and there would be a number of implementation difficulties. For example, an 'anyone out there' message could not be sent on the same frequency as the serving cell because it would cause interference and might not be heard in any case, since the signal from the serving cell might interfere with it. Therefore, a separate channel would need to be defined and mobiles would need to monitor both frequencies.

A simpler technical solution might be to use a completely different wireless channel – for example WiFi. This overcomes many of the problems identified above, although it still requires the forwarding mobile to monitor WiFi channels frequently. Further, there may well be WiFi coverage within the building, in which case it would be simpler for the mobile to make direct use of that channel.

Another problem with using other nodes for forwarding is the user's acceptance of this. Forwarding will use battery power and mobile owners might decide that they do not want their battery drained by forwarding messages for others. There might also be concerns over security and privacy.

In a disaster where the conventional networks have failed, the situation is somewhat simpler. In this case there is no coverage, so anything that mesh can provide is an improvement. There are no networks to interfere with. Devices that can revert to 'mesh mode' when all else has failed might have value. To some extent this is already possible. WiFi has a mesh mode and many handsets now have inbuilt WiFi radios. Better solutions using more advanced mesh protocols and more appropriate (lower) frequencies could be imagined, but the cost of building these into handsets is probably not worthwhile given the relative infrequency of complete network failure.

Overall, mesh does not appear to be a good solution to coverage. Only where there is no other coverage might mesh systems provide a 'second-best' alternative, but the situations where this might be valuable seem rare.

5.5 Problems with mesh systems

The biggest problem with mesh systems is a lack of reliability. With many mesh systems there is no guarantee that a packet of data will arrive at its destination, or, if it does so, how long it will take. For many applications this lack of certainty makes mesh systems unusable.

The problem arises because in a classic mesh system the route that the packet will take from transmitter to sink node (or end recipient) is not known at the time of transmission. When a packet arrives at each node en route, the node uses its routing algorithms and a search of other nodes in the vicinity to determine where next to send the packet. Because nodes may be in motion, the topography of the mesh is continually changing and the optimum route may change from packet to packet. It is even possible that there will not be any other nodes in range and hence the packet cannot be forwarded, at least until nodes appear.

The simplest solution to this problem is to use a static mesh network. If the nodes are unchanging, or rarely changing, then routes through the network can be pre-determined. If necessary, 'holes' in the mesh can be filled with seed nodes. This was the approach adopted, for example, by Radiant in their design of a wireless broadband to the home system based on mesh. As each home registered for service propagation, software was used to determine whether it would be able to see other nodes in the network. If it could not, then seed nodes were considered in order to enable service. While requiring the mesh to be static might seem a major disadvantage, there are many systems where static nodes occur. These include many sensor networks, home broadband networks and in-building networks. We consider the use of mesh technology to enable these systems in Chapter 13.

5.6 Verdict

Mesh technology has been a topic of much interest for over a decade. Many companies have been started to exploit mesh solutions, but few exist today. Many mesh-based network ideas have been trialled, but most have failed. Practice has also shown that many of the theoretical promises of mesh cannot be realised in the real world. It is clear now that wireless mesh networks are not a part of the mainstream wireless solution for cellular, broadcasting or other similar systems.

Mesh does have a role in specific niches. These are where most or all of the following apply:

- the nodes are stationary or move rarely;
- delays in data transmission can be tolerated;
- nodes are all under the control of the same company or user;
- extremely low-power operation is needed.

Mesh is finding a role in sensor networks, home entertainment systems and military applications. Further development and trials are needed, but it seems clear that it can make a significant difference to a number of applications in these narrow areas.

Reference

[1] S. Methley, *Essentials of Wireless Mesh Networking*, Cambridge University Press, 2009.

6

Software-defined radios and new receiver architectures

6.1 The software-defined handset

Mobile phones are becoming increasingly complex. Over time they have added both additional cellular frequencies and standards and also other wireless technologies. A high-specification phone might now support 2G at 900 MHz, 1800 MHz and 1900 MHz, 3G at 2.1 GHz, Bluetooth and WiFi (both at 2.4 GHz) and GPS (at 1.4 GHz). Some even support mobile TV (DVB-H) at about 700 MHz. Phones in future years will need to support additional technologies and frequency bands.

The traditional way to design a phone supporting multiple standards has been to add an additional chipset to the phone for each standard, although, as time has progressed, some chipset vendors have developed single chips combining some of the most popular combinations of standards. This has the advantage of being relatively simple but the disadvantage of proliferating chips with associated cost. The radio-frequency (RF) design of the phone also becomes ever more complicated. As more frequency bands are supported, the possibility of interference between then increases and the difficulty of designing antennas and filters that can isolate multiple bands grows. Some have suggested that the growth in complexity might be closer to exponential than linear since each new band introduced has to be designed to work with all the other bands already used by the device. To date, these problems have been soluble, with the costs absorbed in the large quantities of handsets sold and with a degree of integration of multiple standards into one chipset. However, as devices become ever more complex there may come a point in time at which a different approach is required.

The alternative is to have a reduced number of multi-purpose systems. This, of course, is the approach taken for PCs, where the vast majority of functionality is implemented in software. Hence the idea that there might be a 'software-defined radio' (SDR).

At its most extreme a SDR would consist of an antenna connected to a very fast A–D converter. This would digitise the entire radio spectrum (say, from 100 MHz to 10 GHz). Once available in digital form, all required operations such as filtering, down-conversion and subsequent manipulation could occur using software. An ideal phone would need only a single (very-broadband) antenna and would be capable of being upgraded to accommodate any new technologies or standards that might be deployed. There would be enormous economies of scale since only one solution would need to be developed for the entire world.

Most accept that such an 'ideal SDR' is unrealisable at the moment and may remain so for many decades. The biggest problem is designing a fast enough A–D convertor with a sufficiently large number of bits that the differences in signal strength across such a broad frequency band can be accommodated. We are still many orders of magnitude away from being able to realise such a device. Producing a wideband antenna is also difficult, especially for small devices, although compromises on its frequency range might be acceptable.

Even though an ideal SDR might not be practical, it is possible to increase the amount of software processing in a handset. For example, most wireless standards require large elements of computation effort in their decoding, so a handset could have a single processing element that was used by all the different standards. Similarly, memory could be shared between standards and possibly some more specialised elements such as turbo-code decoders or fast-Fourier-transform (FFT) elements. The net result might be a handset where the RF elements were bespoke to the various frequency bands used but most of the baseband elements were common, running a different software load depending on the standard in use at the time. This would be akin to loading Microsoft Excel or Word depending on the type of document to be viewed. Some suggest that, with the advent of LTE, we will reach a tipping point where it becomes more cost-effective to build a generic baseband and define each standard in software than it is to incorporate separate chips for each standard. Whether this is true remains to be seen, but it does seem likely that, as ever more standards are supported and each becomes ever more complex, integration of the baseband functions is likely to become increasingly beneficial.

To some degree, this would not change much in the greater scheme of things. Regardless of the implementation approach adopted, the handset would have similar functionality and appearance. Only for the chipset community and handset developers might it be a significant change.

Some have suggested that an advantage of such an approach would be that a handset could download the instruction set for a new standard. This might be because it was in a country using an unfamiliar standard or because a new standard had been deployed. Both of these seem unlikely.

As standards become increasingly global, the chances of a country having its own standard decrease. In 3G there are some regional differences, for example with China adopting its own standard, but this difference is unlikely to be repeated in 4G and in any case there will be coverage from wideband code-division multiple-access (W-CDMA) 3G alongside local standards. New standards are typically about five times more complex than the ones they replace, making it likely that a handset would not have sufficient processing power to operate a new standard. Further, given that standards take up to a decade to deploy while handsets are replaced on a 1–2 year cycle, the new standard can be implemented in the handset prior to it being needed.

Having a software implementation does allow bugs to be fixed through software updates – as is already common with many devices from computers to MP3 players. This is a useful improvement, but not one that materially changes the future of wireless.

In summary, a full SDR is still decades from reality. We can expect an increasing use of generic processors within phones, with more of the standards being implemented in software, but this will not materially change their useful capabilities. It will, however, disrupt the silicon supply-chain and developer community.

6.2 The software-defined base station

While many of the same arguments for and against SDR hold true for base stations as well as for mobiles, there are some significant differences.

- Base stations have a much longer lifetime than mobiles (often 10–20 years). Hence, there is a much greater chance of standards changing during their lifetime.
- Base stations require much more processing power than handsets (because they have to handle multiple transmission streams) and hence are more likely to make use of field-programmable gate arrays (FPGAs).
- Base stations are made in smaller volumes, hence anything that can enable a single design to be sold across a greater number of markets can bring benefits.

Experience has shown that standards often change during their lifetimes. GSM was repeatedly upgraded, with GPRS and EDGE being two of the most significant additions. 3G has been upgraded to HSDPA and HSPA+. It would be expensive and inconvenient for operators to have to replace their base stations each time there was an upgrade of this sort. Hence, for many years, base stations have had the ability to accommodate software upgrades to a standard they already support. Few, however, have been able to accommodate upgrades to new standards. This is broadly for the same reasons as pertain to handsets, namely that new standards are often more processor intensive and base stations are typically not built with substantial spare processor capacity since that would add significantly to the cost of the base station, with no guarantee that it would be appropriate or sufficient when any new standard is deployed.

There have been some base stations, for example those manufactured by Vanu, that implement the baseband processing entirely in software and can be changed from one standard to another. These are designed to implement a number of existing standards – which is particularly useful in countries where there are multiple different standards in operation. This enables an operator to move from one standard to another depending on changing demand, or to share infrastructure with other operators using different standards. Because each of the standards to be supported is known at the time when the base stations are designed, appropriate computing resources can be added to the base station.

Software-defined solutions are also useful in the early stages of a standard. At this point the standard might not be completely stable, so implementing it in software allows for the possibility of changes needing to be made. Also, while volumes are relatively small, it is less expensive to implement a solution on a general-purpose digital signal processor (DSP) or processor than it is to design and manufacture silicon. As volumes increase and the standard stabilises, manufacturers invest in custom chipsets and place more of the processing in hardware. For example, at the time of writing, many implementations of femtocells were using the software-defined chipset developed by Picochip which allowed rapid implementation of the emerging standards as well as operator-specific implementations where necessary. If the femtocell market grows, it can be expected that manufacturers will migrate to dedicated silicon solutions.

So SDR also has its place in base stations, where it tends to be more prevalent than it is in handsets. As in handsets, it does not make a major difference to the capabilities of the equipment, but does have implications for the supply chain.

6.3 Trade-offs in receiver design

The performance of wireless systems is heavily dependent on the quality of the receivers. If the receiver is not sufficiently sensitive to be able to detect weak signals then the range of the system can be significantly reduced and, if the receiver is poor at filtering signals in adjacent bands, then its performance can be compromised, especially when close to transmitters from other radio systems.

Advances in baseband processing have followed Moore's Law, enabling ever more complex modulation and coding schemes to be used. However, advances in RF components such as filters and low-noise amplifiers have been much slower. There has been some improvement; for example, the sensitivity (the minimum signal level that is needed in order to decode the signal correctly) of GSM handsets increased from about −102 dBm when GSM was first introduced in 1992 to about −107 dBm in 1997 (increasing by about 1 dB/year), but then flattened out at −110 dBm in 2007 as both theoretical and practical limits were reached. This progress is comparatively slower than Moore's Law and has reached its limit. Improvements are becoming progressively harder as more frequency bands are added into phones, requiring greater flexibility, broader filters and RF components to work over a broader frequency range.

Indeed, advances in silicon engineering may work to reduce receiver performance. It is becoming increasingly possible to integrate almost all the RF components onto the chip that performs the baseband processing. This brings important cost-savings, but can reduce the RF performance, since the integrated components do not perform as well as did the discrete items which had previously been used. It is the cost trade-off that lies at the heart of receiver design – it is possible to design excellent receivers or low-cost receivers but not receivers that have both of these characteristics. (There is also some evidence that handset designers in emerging companies, for example in South East Asia, do not have the same RF skills as those in more established companies and are producing poor designs, but this can be expected to improve with time.)

Problems may also occur due to more intense spectrum use. One of the key determinants of RF performance is the first filter in the RF chain (often termed the 'front-end filter'). Ideally, this would allow through all the signals in the frequency band of interest while completely blocking all signals outside this band. Practical filters achieve neither. Because they do not allow through the entire signal in the band of interest, they reduce the sensitivity of the device. As a result of allowing through part of the signals in nearby bands, they increase the interference, especially when the receiver is near a neighbouring transmitter. The exact effect of imperfect filtering depends somewhat on the receiver architecture adopted, of which there are many variants.

Receivers can be made better. By spending more money on the receiver, its performance can normally be improved. For example, many receivers have analogue-to-digital (A–D) conversion at some stage. By using a convertor with more bits, a greater resolution can be achieved, which can be used to allow interfering signals in neighbouring bands that are larger than the wanted signal to be passed through the A–D convertor and then removed using digital filtering techniques. However, more bits on an A–D convertor require more space on a silicon chip. This might add $1–$2 to the cost of a receiver chipset for many common receiver designs. When passed through all aspects of the supply chain, this could lead to $5–$10 extra device cost. For smartphones this might not be unduly problematic, but for low-cost receivers this could materially change the price. Manufacturers are under pressure to produce devices that are 'just good enough' and hence are disinclined to implement better receivers unless there is a clear performance advantage for the end user or standards require them to. In many current situations neither is true – as an example, the iPhone is said to have a less sensitive receiver than many other phones but this is not perceived by users to be a material disadvantage.

6.4 Where next for receivers?

The receiver is a critical component in a wireless system. A small decrease in receiver performance can reduce the range of the system, substantially increasing its cost, or lead to unreliable coverage. Poor filtering can result in systems that are affected by transmissions in neighbouring bands, causing coverage 'holes' in the vicinity of neighbouring transmitters. Receiver design and implementation is a compromise between cost and performance, with cost pressure on device designers tending to result in receiver designs that are less than optimal.

There is little on the horizon that is likely to change receiver performance. Advances in RF components are slow and unlikely to change performance substantially over the coming decade.

Reference

[1] http://www.rttonline.com, where this is discussed under 'Hot Topics' for April 2007 and in the accompanying White Paper.

7

Cognitive or white-space systems

7.1 Unlicensed spectrum is becoming more valuable but more congested

When Ofcom last looked at the value that a country derives from its use of spectrum in 2007, it concluded that unlicensed use delivered only about 1% of the overall value. This was a backwards-looking survey that was based on evidence from previous years. To come to these conclusions, the survey assumed that the main value in unlicensed use was WiFi and that this added value by enabling home owners to avoid wiring their home. At the time and looking at the evidence available for earlier years, this may have been a reasonable characterisation. It led to the overall assessment that spectrum managers should concentrate on licensed bands, where the overwhelming value of the spectrum was to be found.

Since that time much has changed. WiFi has become more than just a wire-replacement technology. It is slowly becoming a core part of our communications network, used by many on a daily basis to improve productivity and access network resources in a range of locations. It is increasingly used by cellular operators as a means of offloading data traffic, to the extent that in a few years time a widespread failure of WiFi networks could result in serious congestion on cellular networks. WiFi or other unlicensed devices may form home networks that can deliver important benefits in terms of energy efficiency, assistance in the home to the elderly and infirm, and a core part of the home-entertainment proposition. WiFi in industry might deliver substantial efficiency savings, such as using WiFi in hospitals to enable access to patient records at the bedside and facilitate better communications between staff. A paper from Perspective (sponsored by Microsoft) [1] sets out some of the possible future benefits.

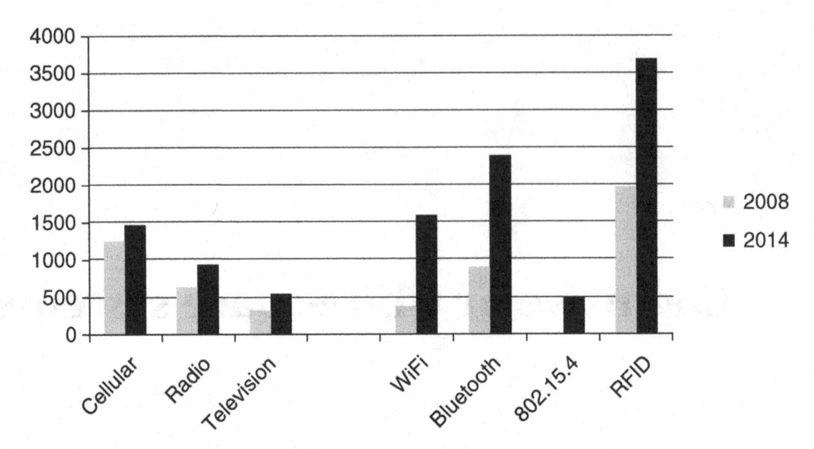

Figure 7.1 Relative volumes of licensed and unlicensed devices (*Source: Perspective*).

For example, the Perspective paper provides Figure 7.1 showing relative volumes of licensed and unlicensed equipment. This predicts relatively slow growth in licensed devices but substantial growth in all categories of unlicensed, to the extent that, for example, WiFi device sales will be greater than sales of cellular devices. This sounds reasonable, since most cellular devices will have WiFi embedded within them and there will be many standalone WiFi devices. Of course, volume of devices does not necessarily correlate with value added; nevertheless these predictions suggest a future in which unlicensed usage becomes an increasingly key part of modern wireless communications.

The paper from Perspective sets out the example of a hospital where WiFi is used and notes the very substantial value that can be generated through efficiency savings. However, if WiFi is used in this manner it could also become an essential element of healthcare delivery, with important implications for reliability and interference. Perspective suggests that unlicensed use may add \$16–\$37 billion annually to the US economy over the next 5–10 years. The Ofcom report estimated the total value from all spectrum uses in 2006 at £42 billion. Scaling the US numbers to the UK by population and adjusting for the exchange rate leads to estimates of unlicensed value being some 5%–10% of total spectrum value over the coming years.[1] Making predictions of this sort is fraught with difficulties, but these suggest that the value of unlicensed spectrum will certainly grow, perhaps by as much as an order of magnitude.

As the use of unlicensed spectrum grows, so does the possibility of congestion. The 2.4-GHz band which is used by most applications is only 83 MHz wide

[1] This is highly approximate, depends on the exchange rate and does not account for the likely growth in value in licensed applications over the period.

and is shared by both communications systems and sources of interference such as microwave ovens. Different communications systems are more likely to interfere with each other than similar ones, further adding to the potential for problems. Indeed, in the past significant interference between WiFi and Bluetooth was experienced, leading to a modification to the Bluetooth standard to facilitate better interworking with WiFi. With no central planning, no guarantees against interference and growing usage, congestion in the band is a serious possibility.

Understanding whether congestion is occurring is very difficult. Unlike in the case of cellular networks, there are no central control systems able to gather information on performance. Drive-tests do not work for WiFi since most of the usage is indoors. Congestion can vary dramatically over a short distance, from high usage within an airport to hardly any in an adjacent car park, and can vary according to the time of day. Many users have noted the availability of a large number of WiFi networks in particular locations and surmised that this leads to congestion, but in practice WiFi networks are generally able to self-coordinate and coexist well such that noticeable congestion does not occur. Anecdotal information is often unreliable – a user may perceive a poor service but this is often due to problems in other parts of the link rather than the wireless network – for example, many WiFi deployments are limited more by backhaul than by the wireless connection. It seems unlikely that there will ever be a clear, comprehensive and exact means of measuring the level of congestion in unlicensed bands.

In 2008 Ofcom sponsored a study on congestion at a number of locations both in London and in various other selected areas in the UK. This was based on measurements from handheld devices, which monitored details of the WiFi signal such as signalling frames in order to understand the effect of any congestion on user perception. While noting many difficulties with accurately characterising user perception, the authors of this study concluded that interference from devices other than WiFi was far more likely to lead to user problems than was congestion caused by too many WiFi networks. Indeed, it was noted that WiFi was extremely robust with respect to high-intensity WiFi usage in the vicinity. This correlates with earlier experience of Bluetooth usage leading to WiFi problems. The Ofcom study in particular highlighted video senders as a likely key source of interference. These are devices that re-broadcast a TV signal from the set-top box normally located alongside the main TV to other TVs within the home. The broad bandwidth and continuous transmission from such devices make avoidance of interference particularly problematic. Congestion problems were noted throughout central London.

Unlicensed transmission is uncertain. Congestion or interference can occur at any point – for example, a network that was working well might suddenly be disrupted by someone nearby turning on a video sender. There would generally be no recourse for the WiFi network owner and limited options to resolve the problem. WiFi might become a victim of its own success, with the very characteristics of unlicensed spectrum, such as free access and the ability to innovate, working against WiFi as it becomes more established and important.

For many users there is no need for a solution. Many systems in the home work well and the likelihood of interference from neighbours may be low, especially where homes are relatively well spaced apart.

For larger offices and public buildings the solution may lie in managing the unlicensed usage in their premises. For example, in the new Terminal 5 at Heathrow detailed spectrum planning took place to ensure that interference did not occur among all the different uses of unlicensed spectrum. Businesses within Terminal 5 are not allowed to deploy their own networks without approval from the internal spectrum-planning authority. Businesses can similarly place some control over the use of spectrum within their buildings (although they can do little if significant interference arises from external sources). This may be a viable solution for the case of hospitals highlighted earlier, which already place restrictions on the use of mobile phones within their premises.

It is also possible that the industry itself may provide a solution. Manufacturers of video senders have already noted that many of their customers also wish to run WiFi networks in their home and might return the video sender if it caused unacceptable interference. As a result video senders advertised as 'WiFi friendly' have started to appear on the market. While it is not clear how the 'friendliness' is being achieved, this does indicate that industry is seeking solutions to these problems and increases awareness among individuals that they might need to perform some rudimentary interference management within their homes. As WiFi becomes more valuable, devices that cause interference will become less attractive to many possible purchasers, with the result that manufacturers may seek alternative spectrum or robust mechanisms that prevent interference. None of this, though, is certain.

The most certain solutions are regulatory ones. Regulators could provide more spectrum, could tighten the rules of usage of existing spectrum or could move WiFi to a band reserved for WiFi use alone.

More spectrum has already been provided – in most countries there is some 400 MHz of spectrum available for WiFi use in the 5-GHz band. This is about five times as much as at 2.4 GHz. Standards have long been in place for WiFi use at 5 GHz and some equipment has been available. However, industry has generally preferred to stay at 2.4 GHz. This is most probably a reflection of the low cost of 2.4-GHz chipsets and the widespread availability of equipment. It may also reflect the fact that most users are content with the experience of using the 2.4-GHz spectrum and do not see the need for more expensive solutions to resolve congestion problems. Propagation at 5 GHz is worse than at 2.4 GHz, reducing coverage, although it is not clear whether this is a significant problem. So, if interference were experienced at 2.4 GHz, businesses and users could migrate relatively quickly to 5 GHz. Of course, there are no guarantees that interference will not also occur at 5 GHz, but the much broader bandwidth and the lack of devices such as microwave ovens suggests that the probability of interference will be much lower.

The regulator could instead tighten the rules on using key unlicensed bands, particularly the 2.4 GHz band. For example, entry of new devices or technologies could be restricted solely to those for which it could be demonstrated that they would not have deleterious effects on WiFi and other key users of the band. This sounds attractive, but there are a number of issues.

- Any restrictions would probably need to be on a global basis, since devices are highly portable, and this might be difficult to achieve.
- It might already be too late, with video senders already causing interference. Removing devices once they have been purchased is almost impossible.
- It might stifle innovation by preventing new services entering the band. Indeed, perversely, it might prevent evolution of WiFi systems.
- Users of other services both in the 2.4-GHz band and in other bands might claim that they also deserve protection. More generally, it might raise an expectation that services launched in unlicensed spectrum that became successful would gain regulatory protection, which might have unforeseen consequences.

Nevertheless, given the value that WiFi might bring in the future, regulators should give some consideration to entry restrictions designed to preserve key services.

An alternative is to move WiFi to its own dedicated band. This would guarantee that the only interference came from other WiFi use. However, finding such a band on a global basis is very difficult, and, even if one were found, migrating from the current band would take some time. Indeed, arguably, the 5-GHz band is likely to end up being used only for WiFi in any case and hence such a band already mostly exists (but, as mentioned earlier, has not proven popular).

Another source of additional spectrum might be white space, or cognitive access. Some have noted that cognitive access is quite well suited to home WiFi-type usage because a static node in the home can determine the channels in local use relatively easily and then coordinate the other devices in the home. The lower frequencies of the TV spectrum, where cognitive devices are expected to operate, may benefit coverage throughout larger homes (although equally they may lead to more interference between homes). One downside is that the channel bandwidth is lower (6–8 MHz compared with 20 MHz for WiFi), although it may be possible to combine channels or lower data rates may be acceptable, especially given that most WiFi systems offer far higher data rates than that of the backhaul connection to which they are connected. Cognitive systems might be better at avoiding interference from other sources because of the broader bandwidth available to them, and for home-based systems there is the possibility of coordination making use of the database facilities provided for identifying free channels.

This chapter concerns cognitive access and the issues surrounding it.

7.2 Defining cognitive systems

The thinking behind cognitive devices has undergone much evolution over the last five years and hence the terminology has changed as well. The original idea of a cognitive device was one that, when turned on in a new environment, would scan the radio spectrum and make intelligent deductions (hence the use of the term 'cognitive') about which spectrum was being used, which was free and hence which frequencies it might use to make transmissions. It would then adapt its transmission parameters such as bandwidth and modulation type to best fit the available spectrum. One of the initial uses of cognitive devices was for military units in foreign countries. Often the use of spectrum in these countries is little known or there is insufficient time to determine the use and programme all the wireless devices before they are needed. A cognitive device that could automatically find the optimal spectrum and use it to communicate with other devices would be advantageous. While information about military use is not widely available, it does appear that the US military, and perhaps the military in other countries, is pushing ahead with the development and deployment of such cognitive devices.

Once the concept had been defined, many in the commercial sector started to ask whether it could be extended. The idea was that cognitive devices could identify spectrum that was licensed to others but unused by them and temporarily make use of it, expanding significantly the amount of unlicensed spectrum available. As the idea developed further, proponents focussed on the UHF TV spectrum as a first band where the concept might be tried. Maps of TV coverage in the USA tended to colour the parts of the country where there was TV coverage and leave other areas white. As a result, cognitive devices intended for commercial use in the TV bands came to be known as white-space devices.

While these white spaces extended across the largest expanses of spectrum in rural areas, they also tended to be quite extensive in urban areas because of the need to leave channels adjacent to TV broadcasts free of other TV broadcasts.

This chapter first of all discusses the issue of unused spectrum, setting out the rationale for cognitive access but warning that identifying spectrum usage is highly problematic. It looks at the problem of detecting signals and concludes that commercial devices are unable to do this at present, especially for systems with difficult-to-detect signals such as spread spectrum. It then sets out the alternative approach currently being studied, namely that of using databases that can provide details of channel availability in a given location. It then discusses the applications that cognitive devices might be suited for and considers whether the database concept might evolve into a more general spectrum-management tool, before delivering a verdict on cognitive devices.

7.3 The apparently unused spectrum

The reasoning behind cognitive or 'white-space' devices is relatively simple. Scans of spectrum utilisation consistently show that less than 20% of the spectrum appears to be utilised. Most of these scans were made in fixed locations, often in major cities in the UK or USA, but more recently Ofcom has completed a scan covering most of the UK. This is shown in Figure 7.2.

Clear areas of high utilisation exist around the sound and TV broadcasting bands (100–200 MHz and 450–850 MHz, respectively) and the cellular bands (850–950 MHz, 1,800–2,000 MHz and 2,100–2,200 MHz), but elsewhere the utilisation of the spectrum generally appears low.

Given that spectrum is often perceived to be scarce, proponents of cognitive devices argue that, if a way could be found to use this apparently under-utilised spectrum, substantial benefits could result. In the longer term it might be possible to 're-purpose' the spectrum by moving it from its existing application to one, like cellular, that makes more efficient use of the spectrum, but experience has shown that such re-purposing can take a decade or more. In the shorter term, proponents suggested that it was better to build devices that could opportunistically access the spectrum that they find unused in their location.

However, utilisation charts such as Figure 7.2 need to be treated with caution. There are many uses that measurement systems will not detect, including the following.

- Satellite transmissions, which are typically too weak to be detected unless special equipment is incorporated.

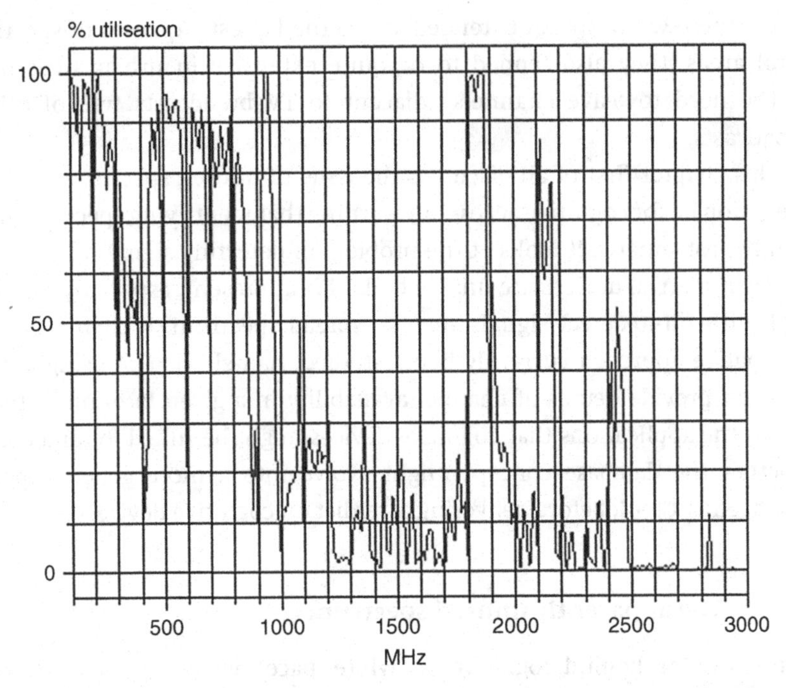

Figure 7.2 Utilisation of the UK spectrum, according to measurements by Ofcom.

- Radar transmissions, for which the pulsed nature of the transmission tends to be difficult to detect with scanning receivers.
- Passive bands, such as those used for radio astronomy or Earth observation, where there are no emissions to detect in any case.
- Mobile uplink transmission, such as from a cellular handset to a base station, where the low height of both the mobile and the detection equipment tends to hugely attenuate the received signal unless the two devices are in mutual proximity.
- Unlicensed bands, where much of the use is at low power and indoors, resulting in weak signals at the detector.
- Much military usage, which may be deliberately covert, or may include radars, aeronautical communications or other difficult-to-detect transmissions.

Hence, utilisation will always be higher than that suggested by measurement systems such as the one used above. Quite how much higher is difficult to say and depends on the precise frequency bands and location as well as the nature of any transmissions to be made.

There is a further complexity in determining how much spectrum is free in that spectrum that is available only in a small area is generally of little value.

Most uses of spectrum envisage a transmitter sending information to receivers away some distance. To be useful, spectrum needs to be available over at least the coverage area of the radio system and in practice some distance beyond this, since the transmitted signal can often interfere beyond the boundaries of its intended coverage. Some of the spectrum identified as unused in these measurements will be in use in nearby locations. This is, for example, often the case with cellular systems that leave some frequencies unused in some locations in order to avoid interference, but reuse these frequencies nearby.

Ideally, the spectrum-utilisation chart shown above would be presented as a 'spectrum-availability chart' showing how much spectrum was available for a given application, taking all the above factors into account. Unfortunately, doing so is extremely difficult and it may be some time, if ever, before such charts are developed.

The most extensive surveys and models have typically been in the UHF TV bands, where about 30%–50% of the spectrum appears to be available to low-power users in most locations. This UHF band has been the focus of work on cognitive devices for a number of reasons, including the following.

- There appears to be a relatively large amount of white space.
- The low frequency of the UHF band results in favourable propagation.
- Licenses are often provided on a single-transmitter basis rather than across the entire country. Therefore, licensed users do not own the white space, unlike, for example, in the case of cellular bands, for which operators typically have nationwide licenses. This means that no negotiation with the licensed user is required in order to secure access; instead, the regulator can legislate to allow it.
- Licensed use is relatively static. TV transmitters are seldom moved. It was also originally thought that the transmissions might be easy to detect, although subsequent work (see below) has cast doubt on this.

Despite all these caveats, it is nevertheless clear that there is some 'white space' even though it is difficult to quantify how much. The next section considers the problem of determining where it is.

7.4 The detection problem

The original proposals for cognitive devices envisaged that the devices would autonomously determine the location of the available spectrum by scanning the spectrum. Where no transmissions were detected, the cognitive device might assume that the spectrum was unused.

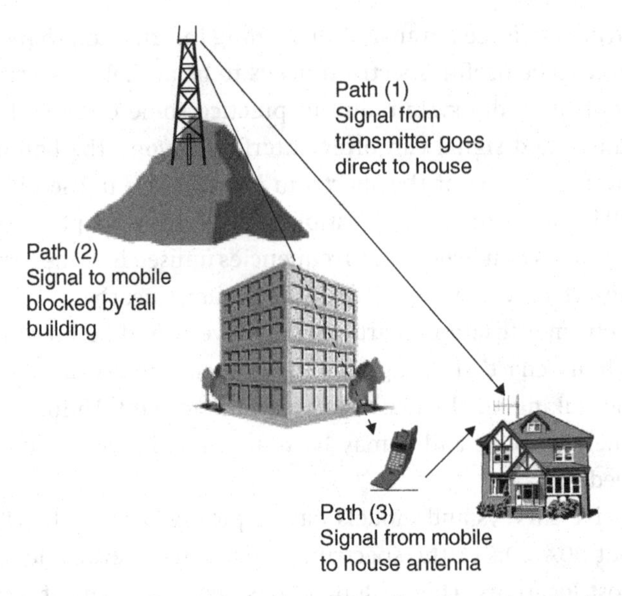

Figure 7.3 The hidden-node problem (based on a drawing by Ofcom).

However, there is a problem with this approach, as was introduced above when discussing measurement systems, which is often referred to as the 'hidden-node problem'. This is illustrated in Figure 7.3. Essentially, the problem is that the cognitive device might not be able to detect a signal because of local topography and hence may transmit in error, causing interference.

Resolving this problem directly either requires the primary user of the spectrum to accept occasional interference, or requires the cognitive device to be sufficiently sensitive that it can detect the signal even when the topography is problematic.

In general, primary users are not inclined to accept interference. For example, TV broadcasters argue that their viewers will not tolerate occasional loss of signal while radio-microphone operators point out the highly problematic consequences of a radio microphone suffering interference during a major stage show. There may be some users who are less concerned about interference, but these have not emerged to date.

The alternative is to make the cognitive device sufficiently sensitive that the hidden-terminal problem occurs so rarely as to be considered resolved. In the USA and UK substantial work has been undertaken to determine what sensitivity might be needed. Definitively determining sensitivity is not possible because it depends on real-world geometries, obstacles and topographies, but substantial measurement and modelling have provided good guidance. The answer is that cognitive devices need to be extraordinarily sensitive, so much so that it is unclear whether such devices can actually be realised, let alone at a price that would appeal to consumers. It is possible that novel algorithms that can measure well below the noise floor might be found in the future, but for the moment most accept that cognitive devices are unlikely to be based on detection.

7.5 Databases as an alternative

If the cognitive device cannot determine accurately the spectrum available, then an alternative is for a central infrastructure to carry out this determination and transmit the information to the device. In such an approach, a cognitive devices determines its location, perhaps using GPS, and then reports this, perhaps via a cellular data channel, to a central database. The database returns information on the frequencies available in the vicinity, possibly including additional information such as the transmit power that the device might employ or the likely duration for which the channel will be available.

For such an approach to work, the database must be supplied with detailed information about the licensed use, including transmitter locations, transmitter power, antenna orientation and so on. It must also be aware of the operating parameters of the licensed systems such as the carrier-to-interference (C/I) level that the service needs in order for it to operate successfully. With this information it can derive a model of possible receiver locations and signal strengths and then, for every given point on a map, it can determine what signal strength a cognitive device could transmit without causing interference with the licensed service.

While this resolves the problems with detection, it requires considerable organisation and agreement. One or more databases must be created and maintained, and protocols designed for communicating with them. The cognitive devices now need to have a means of determining their location and also of communicating to the database using non-cognitive wired or wireless systems. In addition, licensed users need to update the database whenever their use changes, which can be frequently in the case of uses such as wireless microphones and cameras. Most imagine that access to the spectrum by cognitive devices will be free, and therefore the sources of revenue to enable such deployments are unclear. Whether commercial arrangements and international harmonisation to enable databases can be found remains to be seen.

7.6 Applications

The assessment of the amount of white space and the development of solutions to utilise it could be considered as something of a 'solution in search of a problem'. The solution was a way to access additional spectrum, but the need for that access was not clear. In this section we discuss what applications a cognitive device might be suited for.

Cognitive access is somewhat uncertain. In any given location there may be more or less spectrum available, conceivably even none. Available spectrum must be shared with other cognitive users, with the result that congestion might occur. Alternatively, the licensed user might change their usage, instantly removing available channels. Such uncertain access often discourages network deployment because of the risk that the significant capital expenditure needed might not be recovered if spectrum availability declines. Further, it is difficult to offer any kind of quality-of-service guarantees when the access to the spectrum itself is so uncertain. This suggests that cognitive access is not likely to be used for applications such as cellular or broadcast networks. Transmit powers also need to be relatively low in order to avoid interference with licensed users some distance away – in the UHF TV bands transmit powers of less than 20 dBm (100 mW) have been suggested. These are at about the level of handsets or unlicensed devices, but well below those for base stations (typically 10–1,000 W) or broadcast systems (often 100 kW or more).

Instead, cognitive access appears to be better suited for unlicensed applications such as those provided by WiFi, Bluetooth and other short-range solutions. In comparison with existing unlicensed systems, cognitive devices have the following advantages.

- Potential access to more spectrum. Technologies like Bluetooth operate in the 2.4-GHz band where there is 83 MHz of spectrum. In the UHF band alone, measurements suggest that there might be about 150 MHz of available spectrum, and more might be found in other bands.
- Access to 'better' spectrum. Signals propagate further at lower frequencies, or can be made to propagate as far with less transmitter power. Both of these effects can be highly beneficial to unlicensed use, although in dense areas better signal propagation can sometimes simply result in greater interference rather than an enhanced service.

However, there are many disadvantages.

- Devices are substantially more complex than WiFi or Bluetooth. They need location systems and alternative communications channels as well as substantial frequency agility. Unlicensed devices are generally of low cost and adding this complexity may make them overly expensive as well as reducing their battery life. There are some possible solutions. For example, smartphones already have location capability and alternative communications channels and might be able to act as a local 'master' informing nearby devices of channel availability without them needing to consult a database directly. Nevertheless, cognitive solutions will compete with existing unlicensed technologies and any increase in price will need to be justified.

- The start-up time may be longer. Finding other cognitive devices in the vicinity is a complex process, given the wide number of channels over which they might be available. Researchers are only now turning their attention to the problem of how to establish cognitive 'nets' of communicating devices and how to maintain these as the licensed use of the channel changes. In some cases, devices might be able to 'learn' about the channel which they should use and default to it quickly.

Proponents of cognitive devices have taken two different approaches to discussing applications. The first is the 'build it and they will come' approach. Proponents of this approach suggest that predicting applications is prone to failure, but history suggests that, if spectrum is made open and available, then novel applications will emerge to use it. They cite WiFi and Bluetooth as examples of this approach. The second approach is to try to guess a starting list of applications, noting that others may emerge. This list of applications includes the following.

- Rural broadband. Spectrum tends to be less used in rural areas, while broadband coverage (e.g. from cable networks) tends to be sparse because of the cost of roll-out. Proponents suggest that wireless could provide a solution. In practice, this is not a spectrum issue but one of economics. If it were economic to provide broadband coverage then this would have been built by cellular operators or other owners of spectrum. Only if subsidies are provided will rural broadband become economic, at which point licensed solutions are likely to be just as good as, if not better than, cognitive ones.
- WiFi range extension. WiFi systems are often used to provide wireless coverage throughout a house, but the WiFi range is insufficient for some homes, especially if the base station is not sited centrally. A system at lower frequencies would provide better coverage. However, it might also result in more interference if the signal from neighbouring homes propagated further.
- WiFi capacity enhancement. WiFi systems at 2.4 GHz are now becoming congested in some urban areas. Additional spectrum

provided by cognitive access could ease this. However, there is additional spectrum already available at 5 GHz, which does not require the complex cognitive access, although the range is lower.

A key question for cognitive access is whether the additional spectrum access enabled is sufficiently valuable to justify the extra costs and complexities of a cognitive device compared with current unlicensed technologies. This is still very much an open question. Proponents of cognitive access include powerful companies such as Google, Microsoft and Dell, which are sufficiently convinced of its benefits to invest significantly in developing prototypes and lobbying regulators.

7.7 Longer term – databases as a spectrum-management solution

The current focus for database cognitive solutions is the UHF TV bands, for some of the reasons set out above. However, there is no reason why the database should not be extended further. In principle, it could be made to cover all useful spectrum and become the default manner in which spectrum is managed. This might have significant implications for wireless communications.

Once a database with its corresponding protocols has been established, extending it to other frequencies would be technically simple. All that would be required is for other licence holders to supply their usage information.

One issue is more of a philosophical one. As mentioned above, many licence holders have nationwide licences. It is arguable that any white spaces in their frequency bands belong to them, to dispose of or use as they see fit. A licence holder might argue that they should be able to profit from the use of their white spaces, just as a home owner would expect to profit from the occupancy of their home while they were away for an extended period. Alternatively, the regulator (or others) might argue that, if the spectrum is unused, then there is no cost to the licence holder in others accessing it, and that this would provide benefit to the country. There is no right answer to this issue, but it might be expected that licence holders would object to and perhaps legally challenge unlicensed access in their bands unless they were able to sanction it and profit appropriately.

Assuming that this issue can be resolved – either via regulatory action or by the development of an appropriate marketplace for cognitive access that allowed for payments for the use of the spectrum – then there will be some types of usage of spectrum bands in which cognitive access is more or less appropriate. Some examples are listed below.

- Broadcasting. Cognitive should work well in broadcast bands since the transmitters are static and easily characterised and their coverage is well understood.
- Cellular. There is relatively little white space in cellular networks, other than in rural areas, and the network configuration can change relatively quickly. This might be a poor candidate for cognitive access.
- Military. While the military use of spectrum is typically classified, it is well known that there are substantial white spaces, often in the attractive urban areas. Military spectrum managers may be disinclined to sell their spectrum in case they have a future need for it, but might allow cognitive access and might not expect to profit from this. With bands spanning much of the spectrum, access to military frequencies would seem a promising area for cognitive radio.
- Aviation/radar. Aviation uses a large percentage of the lower-frequency spectrum, mostly for radar systems. These have many similarities to broadcast systems with a few static locations transmitting high-power signals. However, radar systems have safety-of-life functionality such that interference can be highly problematic. This might be a promising area for cognitive access, but only if it can be clearly demonstrated that access is possible without in any way endangering the functionality of the radars.
- Unlicensed bands. There is no need for cognitive access to unlicensed bands. Access is allowed on a simpler basis already.
- Satellite. For uplink bands cognitive access might be possible in areas away from the Earth station. However, for downlink usage, unless the location of the receivers is well understood, there would seem to be a high probability of interference.

Database access could, in principle, be taken even further. Licensed use could be removed from multiple frequency bands and all access could be via a database. This is the 'spectrum commons' utopia advocated by some.

However, many applications require certainty of access. Few operators will invest in costly widespread infrastructures if they are uncertain as to whether they will be able to access spectrum once systems have been deployed, or whether interference might reduce the capacity of their network. A system of cognitive access without any ability to reserve spectrum might discourage investment. Widespread cognitive access would also complicate devices such as broadcast receivers, which would have to scan across a wider range of frequency bands in order to determine where transmissions were currently located, while safety-of-life systems such as radars would not operate well in an environment where access to interference-free spectrum was not guaranteed. Making all spectrum users adopt cognitive access on an unreserved basis does not appear appropriate or likely. A reservation capability could be built into any database, but if long-term reservations were made then this would essentially be the same as a licence for the spectrum.

Extending the database to other frequency bands would seem appropriate; moving to a method of using the database as the primary method to manage radio spectrum would not.

7.8 Verdict

Cognitive access is less a new technology than a new approach. Cognitive devices, probably using database access, need to implement a location sensor (e.g. GPS), a communications channel (e.g. GPRS) and a mechanism for tuning across a range of frequencies according to the instructions received (as many receivers, such as those in cellular handsets, already do) – but none of these require new technology. Only if detection is to be used rather than databases might novel technical approaches to the problems of detecting signals at very low levels be needed. Instead, a new approach to managing devices and regulating spectrum access must be adopted in order to implement and update the database. This is clearly possible technically and procedurally, but might take some time and harmonisation before it can come into practice.

Cognitive applications are not easily predictable, but seem likely to be mostly restricted to unlicensed access, where the market already has multiple successful technologies to select from. At best, cognitive access might facilitate a range of new unforeseen applications that benefit from the greater range at lower frequencies or the access to more spectrum. At worst, the added complexity of cognitive access might increase the cost of devices to a level at which they cannot compete with existing unlicensed solutions. It is clearly worth facilitating cognitive access to see which of these outcomes will transpire.

References

[1] http://www.ingeniousmedia.co.uk/websitefiles/Value_of_unlicensed_-_website_-_FINAL.pdf - accessed Oct 09.

[2] http://www.ofcom.org.uk/research/technology/research/exempt/wifi/wfiutilisation.pdf.

8

Codecs and compression

8.1 Basics of compression

Capacity on wireless channels is often scarce. As discussed in earlier chapters, the capacity of a wireless system is determined by the efficiency of the technology, the amount of spectrum and the number of cells. Adding additional capacity almost invariably comes at a cost and hence there is an incentive to reduce the capacity requirements as much as possible. Only in certain networks, such as within the home, where the wireless systems provide more than sufficient capacity, is this not true – but even in these cases the demands have a tendency to grow to take up the available capacity.

One approach to reducing capacity needs is compression. Many of the types of information to be transmitted have significant redundancy within them – for example, much of speech is silence (between words or when the other party is talking), while one video frame tends to be very similar to the preceding one. In some cases huge reductions in data rates can be achieved by compressing the data stream. For example, one of the major gains in the number of voice calls that could be handled on moving to 2G cellular was the ability to use digital encoders to digitise voice and in the process substantially reduce the data rate needed. In this chapter we cover the current capabilities and likely future progress of encoders and decoders (collectively known as 'codecs') and consider the implications for wireless data requirements in the future.

In outline, compression works through the removal of material that will not unduly affect the quality of the received and reconstructed signal. This might be due to a number of reasons.

1. The material is not needed – this is true of areas such as gaps between speech.

2. Loss of the material will not be noticeable. The eye and ear will often not notice the removal of particular information, or will be relatively insensitive to it, and hence compression systems can either remove it, or reduce the data rate associated with that particular type of information.
3. Lower resolution can be tolerated, perhaps because video is being viewed on a smaller screen.
4. There is redundancy within the material; as mentioned above, one of the most obvious examples is the similarity between most video frames, requiring only the differences to be transmitted rather than each frame in full.

Compression can be applied to any data stream. To make the best use of prior knowledge regarding factors such as sensitivity to information removal, specific codecs have tended to evolve for particular applications. The main types of codecs are for video (e.g. MPEG2), speech (e.g. GSM full rate), music (e.g. MP3) and still pictures (e.g. JPEG).

Compression of a general stream of data is more difficult. In this case the codec must assume that the data has to arrive without any change from that transmitted. Only so-called lossless encoders can be used in this case. One of the best-known examples of this is the 'zip' encoder used on PCs to reduce the size of large files by removing redundant information. For example, this might replace a sequence of n consecutive 0s with the data '$n \times 0$'. However, as many will have experienced, 'zipping' a PC file tends to reduce the data rate no more than to perhaps 70%–80% of the original.

In this chapter we will consider the two major codecs associated with wireless transmission – audio and video – before looking more generally at data-transmission trends.

8.2 Audio compression gains over time

Audio compression involves either voice or music (although codecs can accommodate both, they tend to be specialised for one or the other). For many years, voice codecs have been of key importance for wireless transmissions. Until recently, voice was by far the greatest traffic load on a cellular network. If codecs could be enhanced to compress voice to half of the previous data rate then the capacity of the cellular network (and hence its ability to generate revenue) would be doubled.

Simple audio compression schemes recognise that voice waveforms over a period of 10 ms or so are relatively unchanging and can be characterised by sinusoidal waveforms at various frequencies. Sending the amplitude and frequency of each of these can be much more efficient than sending the original waveform. More complex speech codecs mimic the biological mechanism with which humans speak. They seek to model the vocal tract of the speaker and then understand the way in which this is 'excited' with bursts of air from the lungs. By sending parameters corresponding to both, the speech can be recreated at the other end. Many codecs exist that utilise some elements of each of these two mechanisms.

At first there were rapid gains in speech-coding technology. The earliest codecs operated at 64 kbits/s, but this was rapidly brought down to 32 kbits/s and then 13 kbits/s for the original GSM codecs. However, after this point advances slowed down. Reducing the speech rate further both required substantial increases in computing power and tended to reduce the quality of the speech in certain situations. An attempt to produce a half-rate codec at 5.6 kbits/s around 1997 was not successful because the quality was not perceived as satisfactory and instead attention was turned to an 'enhanced full-rate' codec operating at 12 kbits/s but providing higher voice quality than the original GSM codec. Since then, the key advance has been in multi-rate codecs that modify the data rate they produce depending on the input. Some sounds require less detailed characterisation than others and so lower data rates can momentarily be used. The adaptive multi-rate (AMR) codec varies its rate between about 5 kbits/s and 12 kbits/s according to the input. Such variation is useful only if any unused data capacity can be made available to others – an approach that is simpler to implement using 3G CDMA technology, which inherently shares interference equally among all users, but can be achieved by making some statistical assumptions on GSM.

Since about 2000 there does not appear to have been any substantial improvement in speech rates. While techniques that deliver voice at data rates as low as 1 kbits/s are known (and may be used in some military applications), voice codecs for cellular systems require peak data rates of about 12 kbits/s. Although it is always possible that a breakthrough new algorithm that will enable a further reduction in data rate will be devised, this appears unlikely given all the research that has been directed into this area. Further, as the percentage of traffic that is voice on a cellular network falls, the relative benefits from better voice compression reduce, encouraging a focus on other areas.

8.3 Video compression gains over time

Video compression offers potentially even greater benefits than audio compression. Uncompressed video streams can require data rates in the region of 1 Gbits/s for high-quality images. However, there is so much redundancy in raw video information that compression ratios in the region of 40:1 are routinely achieved. Initially, video compression allowed the use of DVDs to store films but, with the advent of digital broadcasting, it has enabled multiple digital channels to be broadcast within the bandwidth of a single analogue channel. As well as for broadcast channels, video transmission is also starting to become significant on cellular systems and, given the much larger bandwidth requirements for compressed video compared with compressed voice, improvements in video codecs could be of greater importance in the future than improvements in speech codecs.

Simple video compression approaches first removed material that the viewer would generally not notice and then concentrated on the differences between video frames. More complex codecs look for blocks of material in one frame that have moved in the subsequent frames (e.g. a car moving across the screen) and send information on the block and then subsequently on the manner in which it has moved. There have been proposals for future systems that would characterise objects using an underlying mesh framework and then send information on how the frame is moving and how the 'skin' on top of the frame is adapting as a result. While potentially very efficient, these have proven far too complex to implement to date.

Since about 1994 video codecs have improved substantially. In 1994 broadcast-quality video under MPEG-2 could be coded at about 8 Mbit/s. By 2000 the MPEG-4 standard had reduced the bit rate by 10%–50%, with higher reductions for lower-quality video. By 2005, further advances in MPEG-4 coding using the H.264 standard had reduced the bit rate to 20%–70% of 1994 levels. These figures show that the greatest advances have been made in lower-quality video, where improvements of a factor of three or more have occurred. However, as viewers purchase larger flat screens they not only are becoming increasingly intolerant of poor video quality but also are moving to high-definition (HD) content. At these higher quality levels, codecs have exhibited less improvement, in some cases as little as 30% less data over a decade.

Experts observing trends in video compression have suggested 'laws' predicting how compression might evolve. For example, McCann [1] originally suggested that compression might improve by about 15% a year, although in recent years he and others have noted that this prediction is somewhat optimistic. We consider a number of reasons why this might be the case in the next section.

8.4 Difficulties in introducing new codecs

For all codecs, improvements get steadily more difficult. Initial implementations exploit the 'low-hanging fruit' and remove the major causes of redundancy. Subsequent codecs address ever smaller gains at ever greater complexity.

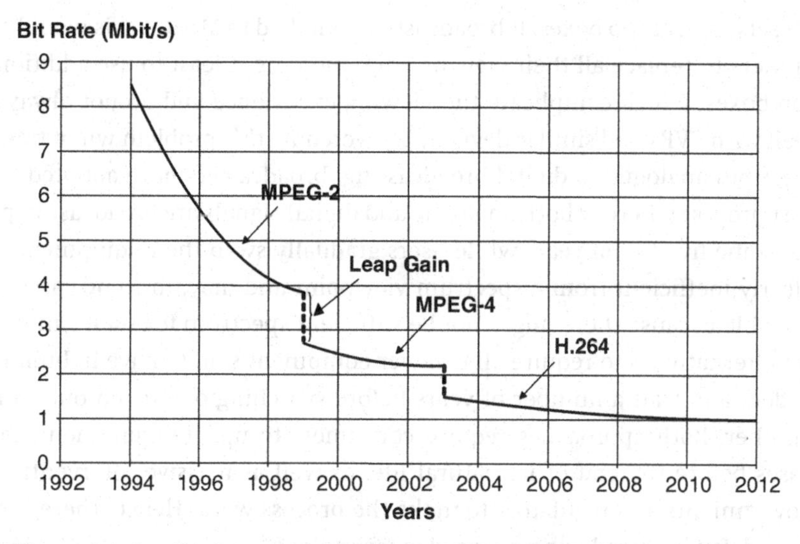

Figure 8.1 A stylized representation of the evolution of the standard codecs over time.

As has occurred with speech codecs, it might be expected that compression rates would fall quickly at first then tend roughly asymptotically towards a final value.

Codecs tend to come in generations. Most compression standards describe the meaning of the data stream produced by the codec but not the mechanism by which it must be derived. This means that it is possible to make improvements after a standard has been deployed as long as the resulting data stream remains within the broad format of the standard. This results in steady improvements once a standard has been launched until eventually the restrictions of the data-stream format prevent further gains and a new standard is required. There is typically an immediate gain when this new standard is introduced as pent-up advances can now be implemented. The cycle then repeats. This is shown in Figure 8.1 for standard-definition (SD) video compression.

Figure 8.1 also shows the overall curve for the totality of all codecs tending towards a minimum level over time. Indeed, there seems to be an assumption that H.264 has derived all the gains that are practical for SD TV and at present there is no further work taking place within standards bodies on new SD codecs. Instead, attention has turned to HD and 3D encoding where gains based on some of the different features of these types of systems may be possible. These different formats will tend to follow similar introduction curves to those shown for SD, although it is likely that they will start further along the curve since much that has been learnt from SD will be applicable in HD and 3D.

The introduction of new codecs tends to be very difficult in many areas. For example, in terrestrial TV broadcasting MPEG2 is currently used in millions of

TV sets and set-top boxes. If broadcasters switched to MPEG4 this would require viewers to replace all their viewing equipment or at least to use additional set-top boxes, which complicate the viewing experience and do not always work well with PVRs and similar devices. To overcome this problem when transitioning from analogue to digital broadcasting, broadcasters have adopted a simulcast process whereby both analogue and digital signals are broadcast in parallel for some five to ten years while users gradually swap their equipment. This is clearly inefficient from a spectrum viewpoint and may, in some cases, not be possible because there might not be sufficient spectrum for such simulcasting. An alternative is to require all receiver equipment sold to have in-built MPEG4 codecs and wait a number of years before switching over from one format to another. Both approaches require consumers to update equipment, in some cases before the end of its natural life, as well as massive intervention from governments and regulators to make the process work. Hence, there is often a long delay in introducing new codec technology.

Only where codecs can be distributed by software, as is the case for new music codecs which can be downloaded to programs such as Windows Media Player, can the introduction of new codecs be performed relatively quickly. Even in this case there may be problems when downloading music to portable devices that do not have the new encoder inbuilt.

Some work performed on behalf of Ofcom [2], taking all these factors into account, suggested that practical video coding rates (as in widespread use) might fall by about 7% per year for SD broadcasting. HD rates might fall less quickly and, indeed, might be offset by ever-rising requirements for higher quality on HD. Rates on Internet Protocol TV (IP-TV) were actually likely to rise both as the availability of broadband to the home improved and as IP-TV material increasingly came to be viewed on a larger TV screen rather than a relatively small computer screen.

We might be at the end of a 'golden age' for video and audio compression. Most of the major gains appear to have been achieved and new codecs are becoming ever harder to introduce. Compression technology appears unlikely to substantially ease future data requirements.

8.5 General data transmission

We have already mentioned that, when the source material is unknown, codecs can only adopt lossless compression, which tends to provide limited gains. The only solution to this is to compress prior to sending the data to the wireless codec. For example, Microsoft Word can be asked to compress embedded pictures to an appropriate resolution prior to saving the document.

Table 8.1 *A comparison of data volumes and airtime consumed for a range of applications*

Application	Data volume	Airtime consumed	Signalling events
Mobile web	1 MB	20 min	100
P2P traffic	1 MB	30 s	1
Email	1 MB	2 h	1,500
Average 3G subscriber	20 MB/day	2,800 min/month	260/day

Some wireless devices can actually make matters worse. Table 8.1 shows that for some applications there are many 'signalling events' whereby the device and the server contact each other frequently to check for new messages or other changes. This can substantially multiply the amount of data transmitted.

It seems likely that substantial improvements can be made for those cases in which there are large numbers of signalling events. By designing the application in such a way that it understands it is using a wireless channel and minimises the resource utilisation, much unnecessary transmission could be avoided.

8.6 Future predictions

Coding of video and audio has been a source of substantial gains in wireless efficiency since the early 1990s. However, we are now seeing diminishing gains and, when this is taken together with the difficulty of introducing new codecs, cannot rely on enhancements in compression to provide significant future capacity enhancements.

The balance of different data types on wireless networks is steadily changing. From being predominantly voice in the 1990s, most wireless networks now carry more data than voice. Some of this data is video, but increasingly the wireless network is unaware of the format of the data that it is carrying. Inefficiencies in the way in which applications make use of the wireless channel are resulting in some of this data traffic consuming more resources than necessary.

Over the coming years we cannot expect to see reductions in compression rates, but we can expect to see applications being better designed to optimise their use of wireless channels, bringing about significant improvements in capacity.

References

[1] ftp://ftp.cordis.europa.eu/pub/ist/docs/ka4/au_ws210906_lepannerer_en.pdf.
[2] http://www.ofcom.org.uk/research/technology/research/ese/video/.

9

Devices

9.1 Introduction

To make use of wireless services, users need devices. These are generally some sort of mobile phone, although increasingly devices include laptop computers with datacards. The capabilities of these devices are an important factor in determining the manner in which they are used and the amount of data they generate – this was clearly shown when the Apple iPhone was introduced and those using it generated some 50 times more data traffic than the average customer.

The phone is an area of intense competition, and innovation and advances in many aspects of its performance can be expected. In this section we consider a number of the key components of the phone, including the screen, input devices, batteries and storage.

9.2 Screens

The screen on a mobile device tends to be a compromise between something small that will allow the device to be portable and something large that will provide a good user experience. Different devices embody different compromises depending on their function and the perceived user preference. (This is less relevant when using a device such as a wirelessly enabled laptop.) Over time, screens have added colour and then become ever more vivid and brighter, with better resolution and more depth of colour. Most screens on mobiles, though, still remain poor for watching video and can be difficult to see in bright daylight.

A key breakthrough in mobile devices would be for the screen not to be constrained by the size of the device. This might be achieved by

- having a foldable or rollable screen,
- projecting a screen from the device onto a nearby surface, or
- making use of other screens in the vicinity.

Foldable screens have been the subject of perhaps the most intense research effort and prototypes have been demonstrated. Such flexible screens require fabrication on a substrate that itself can be flexible, and much of the research has looked at printing screens onto plastic substrates. When such displays might become commercial is very unclear. Some have suggested [1] that flexible screens are still more than a decade away, whereas other manufacturers often seem on the verge of introducing them. Various announcements have been made suggesting that they were imminent, for example in 2004 the BBC reported [2] that Philips had started production, but over five years later no displays are in evidence. In February 2009, HP reported [3] that they were close to making a cost-effective flexible display that also had touch-screen capabilities. In many cases it appears that, while prototypes can be produced, manufacturing flexible displays cost-effectively is still proving very difficult.

Current research is focussed on organic LEDs (OLEDs). These have many desirable characteristics, although their lifetime is somewhat short. Current thinking is that these can be printed onto a glass substrate, allowing conventional manufacturing, but then separated from the glass later in the process and bonded onto plastic. Initial displays may be non-flexible but used in lightweight devices where the removal of the glass provides useful weight savings – e-book readers are often quoted as a possible application. Direct manufacturing onto plastic is still a few years away. Initial screens are also likely to be monochrome and, for devices like mobile phones, users may prefer rigid colour displays to flexible monochrome ones.

Making a call as to when foldable or rollable displays will appear is clearly very difficult. Because prototype devices exist already, it is tempting to assume that they will be in production within a few years, but previous predictions show that manufacturing flexible screens reliably and cheaply is problematic. At any point in time a breakthrough could occur as a result of some innovation either in the screen itself (for example a better OLED) or in the manufacturing process, making flexible screens practical. This may occur within a year or in a decade. It may then take some time for the screens to catch up with rigid screens in terms of brightness, colour, touch-sensitivity and so on. Perhaps it is safest to assume that widespread use of flexible displays is closer to 2020 than 2010.

The idea of projecting from a small device onto a blank surface is an alternative approach. During 2008 the basic components needed to do this started to emerge, with the ability to display at up to 50 inches and able to run for 1–2 hours on a mobile-handset battery. By 2009 the first phone devices with inbuilt projectors were available, although at the time of writing there were very few devices, with these typically being available only in the Far East at about $500 each. Nevertheless, the practicality of a mobile phone–projector has clearly been demonstrated. What is not clear is how widespread they will become. The projector unit adds some bulk to the phone, making it difficult to produce a slim-line unit. Hence, it may be that there is a market segment for a larger smartphone with projector, but that the fashion phone market will opt for a slimmer form factor in preference to a projector feature.

Projection screens would, to some degree, appear to address a different market need from that of flexible displays. Projectors can be used only indoors where a large surface is available – for example, for making presentations or showing pictures to friends. They are not practical for most travel situations. Flexible displays would be useful during travel or when outdoors. The ideal mobile device in the future might incorporate both if there were little penalty in cost, form factor or battery life.

The final possibility for screens is to make use of other, larger screens in the vicinity. Mobiles could connect via Bluetooth or similar to flat-panel displays in the backs of airplane seats, large screens in hotel rooms or other available devices. While technically this is simple, experience with connecting laptops to projection units has shown that there can often be difficulties in practice due to incompatibilities in display settings or just the complexity of setting up the connection in the first place. In situations where the connection can be set up once to use repeatedly, for example in the home, there is probably less need for it since the home-entertainment system might provide a better overall experience in any case.

Overall, we note that the display is an area where advances could make a substantial difference to the usability and experience of mobile devices. Current displays are a compromise between device size and usability. A combination of a rollable display for use when travelling and a projector (or access to other displays) to share material or for use when indoors (but not at home) could significantly enhance a mobile device. While prototypes of flexible devices and early products with projectors are available, we are still some way from large-scale production, particularly of solutions that do not change the mobile form factor while providing the vivid colours and touch-screen capabilities of current displays. Prediction in this area is very difficult, since a breakthrough or innovation could then lead to rapid change, but experience suggests that change is likely to be slow rather than fast.

9.3 Keyboards and other input devices

Input devices are a critical part of mobiles and other computing devices. Over time, the preferred input mechanism has evolved to be a full-sized keyboard and mouse, with speech recognition starting to become more valuable. Full-size keyboards are not possible on mobile devices due to their size, so much ingenuity has been applied to devising alternatives that approach the ideal.

Broadly, the choice for smaller devices has been between a miniaturised keypad and touch-screen operation. Phones such as the Blackberry have trialled a range of small keypads, which tend to work adequately, although they do not allow typing at the same speed as would be possible on a full keypad. Basic phones mimic a keypad using the numeric buttons and word recognition/prediction, which provides acceptable functionality for short messages. Alternatively, a touch screen allows direct selection of links and icons and allows a virtual keypad to be displayed when required. This approach has proved successful in the iPhone, where much of the interaction with the device is to make selections and click (or touch) links. It is generally thought that the virtual keypad is inferior to a miniaturised one for ‘business use’. Touch screens continue to develop, with recent prototypes of screens that can detect the strength of a touch, allowing further degrees of interaction (such as playing a virtual piano with touch representing strength of the key press).

Perhaps the combination of a rollable display and touch screen might provide a better solution. Such a display would make it possible to operate a ‘full-sized’ touch screen or at least to provide a full-sized keyboard, even if there were no tactile feedback from pressing keys. However, unrolling such a display might not always be possible, for example, when travelling, there might not be the space or the supporting structure available to make use of such a keypad.

Around 2005 there were some limited experiments with keyboards that were projected from the device onto a flat surface and then finger touches were detected using the interruption of laser beams. However, this concept does not appear to have made it into production and many difficulties and problems with it can be envisaged.

A completely different approach to input is to use speech recognition, bypassing much of the need for a keyboard. For some time, mobiles have provided voice-activated dialling, including selection from the address book, which has proven mostly successful. More recently, some companies such as SpinVox have offered a voice-to-text service that transcribes voice messages into emails. There has been controversy over the extent to which this is automated or whether human intervention is needed to assist the transcription task. Voice-recognition software has been available on PCs for many years and is steadily improving, although far from widely adopted.

A key problem for voice recognition is that, unless it is close to perfect, errors can be sufficiently annoying that users prefer other mechanisms. For example, wrongly identifying and dialling someone from the address book can be very irritating and more serious errors could be imagined. Given the many different accents, dialects and idiosyncrasies of speech, accurate recognition, particularly without training of the software, is extremely challenging. Achieving better results is likely to require more powerful processing, which is problematic in a battery-powered mobile device because it tends to lead to increased energy consumption. It is possible for the mobile device to send the speech waveform to a server for processing, with the resulting text then being returned (much as SpinVox does), although this requires a wireless link and has some delay associated with it.

Users also appear to prefer not to use voice for many tasks. Navigating around a screen is simpler with a mouse than by talking. In public places, talking may be embarrassing or inappropriate and may require the wearing of a microphone. It is probably of most value when composing longer messages, but in many cases users may prefer to wait until they are near a computer for these.

Voice recognition will undoubtedly continue to slowly evolve as increased computing power is provided and as algorithms improve. It is most successful where the vocabulary can be limited – such as when interacting with a device in a context in which only a limited number of commands make sense. It can be expected to gradually form part of our mechanism for communicating with machines.

In summary, interacting with mobile devices will always be compromised as a result of their form factor. The best solution will depend on what the device is predominantly used for as well as the preference of the user, with the result that there will probably be a wide range of different devices (as we see today). Breakthroughs seem unlikely, although advanced software and applications that make better use of what is already available might make a significant difference.

9.4 Batteries

Battery capacity is a serious constraint for many mobile devices. It prevents high-power processors being used in mobile phones and makes using GPS and WiFi for long periods unattractive. Indeed, battery drain is one of the key reasons cited for not implementing WiFi in some phones.

Battery capacity has grown far less quickly than processing and storage, and much of the gain has been used to build smaller and lighter mobile devices.

For example, it has proven difficult to build laptops with the ability to last a full working day without recharging, especially if wireless networking is being used, unless large 'expansion pack' batteries are employed.

For many years, the future of batteries has been seen to be in rechargeable fuel cells. These have the potential benefit of being refillable and of having higher energy storage than conventional batteries. However, despite intensive research activities over the last decade, fuel cells have not made any substantial impression and are now discussed less than in the past. This is because it has proven difficult to realise practical fuel cells, particularly for small batteries, where the 'engine' of the fuel cell is hard to miniaturise sufficiently. Also, with improvements in power consumption, for the most part rechargeable batteries have proved to be 'good enough', with the convenience that consumers do not need to buy and carry the replacement fuel. Research into batteries has been renewed with the increased interest in electric vehicles. However, these are somewhat different from batteries for handheld devices in terms of their size and current levels, and, even if there were advances as a result, it is not clear whether they would have relevance to mobile devices.

Even if battery technology were to take a leap forwards such that much higher energy storage density became possible, this would bring other problems. The energy density in some batteries is already higher than that in hand grenades and increasing it further would increase the risk of incidents such as batteries bursting into flame. Also, if more power were to be dissipated in mobile devices as a result (rather than the battery just lasting longer) then there would be increasing problems with heat dissipation. Laptops already get uncomfortably hot when used on a lap and further heating would be problematic.

A related area of development is the use of wireless power and wireless charging. The first potentially provides a mechanism to remove the need for batteries altogether. This is more relevant for devices such as sensors that are left in place for many years without maintenance. The use of wireless charging might reduce the implications of low battery life by providing more opportunities to recharge devices during the day.

Wireless recharging is relatively simple if conducted over a short range. The wireless device is placed on a charging pad which might be embedded in a desk or in a car. Short-range inductive coupling can then be used to charge the device. This works by essentially taking the (normally) metallic core that makes up a transformer and splitting it into two parts, placing one in the charging pad and the other in the device. As long as the gap between them is small, a magnetic flux can be developed across the gap with almost the same strength as across a complete core. Many systems have been demonstrated and, indeed, inductive charging is in widespread use in devices such as the electric toothbrush, where there is a need to wash the device and hence a desire to avoid contacts that might corrode or allow the ingress of water. The problem is less about the technology and more about the standard. For inductive coupling to work, appropriate elements must be built into the phone. Arriving at a standard that all manufacturers can agree on and building this into phones is commercially difficult and has been the reason why many start-ups developing inductive-charging solutions have failed to date. It would take the large manufacturers to agree on a standard in the same way as they are agreeing a single standard for conventional phone chargers for there to be success in this area.

More controversial has been the demonstration by MIT of powering a light bulb wirelessly over a range of about 2 m using 'resonant mid-range coupling'. This was achieved using two five-turn coils, one at the transmitter end and one at the receiver end. These were relatively large, with a radius of about 30 cm. The overall efficiency of the power transfer was relatively low, being only about 15%. Further losses in efficiency might occur if multiple devices were powered or if the geometry of the coils were not perfectly arranged.

Such an approach is likely to be inappropriate for handheld devices because of the size of the coils, but could potentially be integrated into devices such as flat-screen TVs. Indeed, at the time of writing a prototype flat screen with wireless powering of this sort was being demonstrated at consumer shows. By removing the need for power cables and linking ancillary set-top boxes and similar devices by WiFi, the need for wiring was removed, making wall-mounting simple.

Clearly, completely wireless appliances have many attractions. However, the low efficiency of power transfer is a significant impediment, especially at a time when consumers are seeking to reduce energy consumption. Practical considerations such as the mounting of the transmitter coils may also cause almost as many difficulties as providing power directly to the appliance would. At present, it is far from clear that this form of wireless power is practical and likely to be deployed.

Overall we do not expect significant changes in battery technology or advances in wireless power. Mobile devices will need to become ever more power efficient and may need to avoid or limit those activities that are associated with a particularly high battery drain.

9.5 Storage

The amount of memory, or storage, available on mobile devices has grown substantially over the last decade. This has enabled devices to take on functions such as those of music players and cameras.

Broadly, storage follows a trend of increasing by approximately an order of magnitude every $6\frac{1}{2}$ years [1]. With the move from hard-disc storage towards flash memory, this has enabled handset storage to rapidly expand towards levels sufficient for most envisaged applications. Within a decade storage will probably be ten times greater, with handsets having 200 GBytes or more of capacity. Hence, unlike batteries, storage will become increasingly less of a constraint over time.

The implications that this might have for wireless transmission are unclear. With more storage, more material can be sent wirelessly and stored on the device. Alternatively, with the ability to store more material locally there might be less need to access material using the wireless network. Greater storage might facilitate more 'side loading', whereby material is downloaded to the home network via the broadband connection and then to the mobile device via the home network. However, this approach cannot be adopted for material that needs to be seen in real time. More storage might place greater pressure on home networks, although other solutions such as wired connections or short-range standards such as UWB might be adopted.

9.6 Additional functionality

Wireless devices have gained ever more functionality over time. Alarm clocks and calculators have been built into phones for years. Music players and cameras are now standard in most phones, and the quality of these has improved to the level that they are serious replacements for dedicated devices. Many phones have GPS location and navigation capabilities and some have compasses in addition to this, so that they are aware of the direction in which they are pointed. Most phones have short-range communications capabilities such as Bluetooth and WiFi. The ability to make and to view videos is increasingly commonplace. This trend seems likely to continue. Phones will replace most electronic devices that we carry and may eventually replace wallets and keys.

Some of the enhancements in functionality and the user interface have been bought about through micro-electronic mechanical systems (MEMS). MEMS are very-small-scale devices fabricated onto the surface of silicon chips using standard chip-fabrication technologies. For example, by etching away material, a mostly unsupported layer of silicon that is sensitive to pressure or movement can be left behind. MEMS devices bring many benefits – they are very small, inexpensive to make and can be fabricated with high precision. They can be used for a wide range of purposes, including as accelerometers, pressure gauges and high-speed mechanical switches.

It is the MEMS accelerometer that has made the greatest change to wireless devices to date. This measures movement along all three axes. It allows the Wii handset to measure how it is moved and it allows mobile handsets to know their orientation. It permits concepts such as shaking the phone to simulate shaking dice. It can couple with GPS to help determine the direction in which a device is being pointed or be used to detect that a handset is now being held to the ear. We may be just at the start of the revolution afforded by MEMS devices such as accelerometers.

MEMS may also have a role to play in RF components. It has proved difficult to achieve good RF design on a silicon chip due to the fact the silicon components do not perform well at RF frequencies. However, using high-speed switches and other novel design concepts has raised the possibilities of moving more of the RF components onto silicon. This may further reduce the cost of wireless devices or extend their capability to handle a wide range of frequencies.

The coupling of wireless communications with other sensors and functionality can be put to interesting uses. One is for mobiles to gather information as they go about their daily movements – for example, they can map the signal strength they receive alongside with GPS data on their location and report this back to the cellular operator, allowing very detailed coverage maps to be built up. As discussed in Chapter 17, information on speed can be used to determine congestion on roads. Volunteers could be given particular sensors, for example to measure air quality, which connected to the mobile perhaps by Bluetooth and provided detailed information throughout a country over time. Mobiles are also increasingly used to provide pictures and videos of incidents that can be used by broadcasters, providing them with a large mobile 'camera-crew' albeit one without training and with a poor-quality camera. Mobiles also locate their users, which allows a 'find my friends' function or means that families are able to locate other family members. A wide range of novel and valuable uses of this sort can be expected to emerge over the coming years as mobiles gain more functionality and it becomes easier to write and deploy 'apps'.

A downside to many of these uses is that the need to frequently power up GPS and other sensors uses substantial power and can significantly shorten battery life. For many users this might not be problematic – requiring them to charge daily rather than weekly – but it does place a limit on what a mobile can be asked to do.

9.7 Usability

None of the components or advances discussed in the previous parts of this chapter are of any value if the device is not easy to use. Probably the greatest breakthrough in recent years in mobile devices is the iPhone – not because of its components but because it provided a user interface that was intuitive, fun and effective (the Nintendo Wii achieved a similar breakthrough in gaming systems). This does not require new technology but rather new thinking. Handset designers need to think carefully and innovatively about how people might use their devices and construct the hardware and software around this.

It is interesting that the breakthrough came from Apple rather than the giants of the handset business such as Nokia and Sony-Ericsson. Perhaps this was because Apple, as a new entrant into the market, could approach it afresh, without pre-conceived ideas as to how a phone should look and be used. It might also have helped that they could construct a new operating system rather than trying to build on legacy platforms such as the Symbian OS developed by Nokia. Perhaps also coming from a background of music devices rather than phones enabled a different mindset.

As phones and related devices gain ever more features, keeping the use of these simple and intuitive will demand ever more ingenuity in design and intelligence in software. This, perhaps, is the single biggest challenge for devices and one where the iPhone has shown the way.

9.8 Verdict

Devices have evolved rapidly over the last few years, with the iPhone providing a significant change in functionality and usability compared with phones of just a year or two earlier. Key breakthroughs that would enable another step-change in functionality include rollable screens and speech recognition, but it appears that we may have to wait until closer to 2020 than 2010 for these to appear.

Battery life will remain a constraint on mobile devices, whereas storage will continue to grow rapidly.

Advances in electronics and economies of scale will enable devices to add ever more functionality and improvements in user-interface design will make them simpler and more fun to use. This is likely to translate into ever greater volumes of data transferred over wireless connections, although this is not certain and some of those wireless connections may be in the home rather than on the cellular network.

References

[1] W. Webb, *Wireless Communications: The Future*, Wiley 2007, Section 6.4.3.
[2] http://news.bbc.co.uk/1/hi/technology/3506289.stm.
[3] http://www.wired.com/gadgetlab/2009/02/you-can-check-o/.

10

Network architectures

10.1 A world of multiple different networks

There is a key trade-off in the design of wireless networks. Lower frequencies are generally preferred for their better propagation, making coverage simpler and cheaper. However, at lower frequencies there is less spectrum available, with the result that networks tend to have a lower capacity. This is a basic law of physics and not something that can be changed in the future with technological advances (although the capacity per unit amount of spectrum does tend to improve over time).

This trade-off means that there is rarely one perfect system. Instead, lower-frequency systems are used to provide coverage across much of the country but with relatively low capacity, while high-frequency systems provide much greater capacity in high-density areas. Hence, we have the current hierarchy of wireless systems, where 2G technology at 900 MHz provides wide-area coverage at data rates of about 100 kbits/s, 3G coverage at 2.1 GHz covers about 30% of a country (depending on the topography and population distribution) at data rates of about 1 Mbits/s and WiFi at 2.4 GHz or 5 GHz covers buildings and very dense urban areas at data rates of about 10 Mbits/s. Such a hierarchy has evolved over time as the most economic way to provide a service as close as possible to the user requirements.

Occasionally manufacturers or research laboratories claim that they have invented a technology that allows both a large range and a high capacity, but invariably these claims prove to be false or overstated.

10.2 Home systems

A natural consequence of hierarchical systems of cells is for there to be very-high-capacity but small cells covering homes (and other buildings). Many homes do now have a WiFi base station providing wireless data coverage throughout the home, although this is less due to some 'masterplan' to deliver a hierarchical network and more to do with the utility of being able to avoid home wiring to link computers and printers and then, more recently, the convenience of being able to use laptop computers anywhere in the home.

There is still much uncertainty as to how home wireless networks will evolve. It is possible that WiFi will become the final solution. Alternatively, other options such as femtocells, UWB, distributed 60-GHz systems and more might sit alongside or replace WiFi.

A key fork in the road is whether voice in the home will continue to be via dedicated home wireless telephones (e.g. DECT), via the mobile device connecting to WiFi or via the mobile device connecting to a femtocell providing cellular coverage. At present, most homes in the UK have a dedicated wireless system, but this is less convenient than making use of the mobile (although it does mean that any of those in the home who do not have a mobile phone still have a means to make calls). Having a dedicated phone also allows access to cheaper call tariffs, although this could also be offered on mobile phones when they are within the 'home zone'. It provides a means of communication when the mobile phone has a flat battery and is a concept very familiar to most families. So it is far from certain that dedicated home wireless systems will be rapidly replaced with solutions that allow the mobile phone to be used, but over time it seems likely that changes in tariffs and culture will make use of the mobile more attractive.

The debate between proponents of WiFi and of femtocells, as discussed earlier in Chapter 3, has been taking place for some time, with no clear resolution in sight. Looked at from afar, it makes little difference which technical standard is adopted as long as there is a wireless node in the home, connected to its own backhaul (the home broadband connection). Any standard will probably provide sufficient connectivity and capacity. The debate revolves around more practical issues. WiFi is in most homes while cellular (e.g. 3G) is in most cellular handsets. Either the home needs to gain a 3G femtocell or handsets need to gain WiFi functionality. Both are eminently possible, and both have advantages and disadvantages.

A key factor in the debate, and a point we will return to later, is the structure of the wireless industry. In many countries cellular handsets are subsidised by cellular operators, who therefore have substantial control over the capabilities and functionality they provide. Many cellular operators are not the same companies as those providing conventional home telephony. It is commercially in the interest of the cellular operator to prefer a femtocell since this

- ensures that customers stay on the cellular network, even in the home, and
- does not open the door to voice over WiFi being used in other environments where cellular calls might otherwise be made.

On the other hand, the home owner might be more inclined towards a WiFi solution for all the reasons set out in Chapter 3.

It is possible that neither WiFi nor femtocells will provide sufficient capacity, or in some cases coverage, within the home. In particular, if HD video is routinely distributed around the home then higher data rates may be needed. In this case, solutions using even higher frequencies and smaller cells may be preferred. Such solutions typically result in cells the size of a room with a wired mechanism used to 'backhaul' data around the house. Possible wireless contenders for very-high-data-rate provision in the home include ultra-wideband and 60-GHz systems. Both of these are capable of many hundreds of Mbits/s, possibly even reaching 1 Gbits/s. However, given the difficulty of distributing data at over 100 Mbits/s around most homes, the problems associated with high-speed wireless and the additional cost of multiple transmitters compared with a single WiFi node, it seems likely that applications will reduce their data-rate requirements in order to be able to work with the simpler solutions.

There does not need to be a single 'winner' – it is quite possible and likely that there will be multiple wireless systems within the home (indeed, there already are with WiFi and DECT cordless systems covering most homes in the UK). However, as one system becomes progressively more established such that many peripherals and home devices are manufactured with that standard embedded, it becomes increasingly harder for other solutions to gain traction.

10.3 Always best connected

The deployment of a hierarchical cell network leads to the obvious implication that users should generally be connected to the smallest cell available to them in their location, with the transition between cells of the same size and layers of cells being handled automatically as they move. Connecting to the smallest cell enables the highest data rates and most efficient use of resources. However, there may be some occasions, such as when moving rapidly, when connecting to the smallest cell is inappropriate because of the very frequent handovers that would be needed.

The concept of being automatically connected to the fastest and cheapest resource is sometimes known as 'always best connected'. While obvious, it is rarely fully enabled in practice. Mostly this is because networks at different points in the hierarchy are owned by different organisations. Cellular operators own the large cells and microcells but individuals or organisations whose main business is not wireless typically own the WiFi hotspots, while individuals own home wireless devices that they backhaul via a fixed line network owned by yet another operator. While there have been many attempts to provide connectivity across all these environments the results have been mixed due to the misalignment of commercial interests of the communications providers involved.

It is the difference between the commercial alignment of the communications providers and the communications needs of users that is currently preventing many obvious improvements in communications systems. As discussed regarding femtocells, it is often not in the interest of the cellular provider for their user to use their home WiFi network for voice communications from their cellular phone, despite this often being what the user would like to do. The net result is that the user is not always best connected and many new services are not provided, or are offered in sub-optimal ways.

10.4 A community network

There has long been interest in the possibility of users getting together and pooling their resources to build a network they can all share. This has a utopian feel to it that pleases some of the pioneers in the Internet and those who feel that large operators do not meet their needs. Many different variants have been proposed, but the basic concept is that each user installs a wireless network in or around their home at their own cost. They then allow others to use their network when near their home as long as reciprocal agreements exist. The hope is that, if everyone in a community joined such a shared network, then there would be good wireless coverage throughout the community without the need for anyone to pay any fees (other than those they already paid for their broadband connection).

Interest in community networks peaked in the early 2000s as a number of players, particularly in the USA, promised to build metropolitan WiFi networks covering entire downtown areas, but within just a few years it had become clear to all players that this was not a commercially viable proposition. Most such networks have now shut down, or have been taken over by the local authorities to be used for emergency services, meter reading and other municipal functions. Essentially, the costs of building a very large number of WiFi cells and backhauling them plus all the subscriber-acquisition costs were greater than the relatively small amount subscribers were prepared to pay for the freedom to use their WiFi devices anywhere. Users also realised that they rarely wanted to use their devices in the park or on the streets and that coverage in coffee shops and a few key buildings was sufficient for most needs. Given this experience, it seems unlikely that metropolitan WiFi networks will be commercially viable in all but the densest of cities.

With the failure of metro-WiFi, interest in community networks has grown again. A number of hotspot providers such as BT (under the name FON) offer home subscribers the chance to join their network by sharing their resources in return for the ability to access other hot spots in the network. Software is provided to partition the home router such that there are no security risks and the amount of broadband resource used by others can be limited by the home owner. Interest generally appears to be growing slowly – for the first few subscribers there are very limited benefits because there are few other hotspots, but as subscriber numbers grow there is a 'network effect' and the benefits grow substantially.

Just how useful these kinds of networks are is not clear. In a small village or town the benefit of WiFi access outside the homes of a few others is probably low. In cities the benefits may grow as the chance of being outside a connected home may be higher and the likelihood of being away from home but wanting a WiFi connection may be greater. Alternatively, where there is a strong community that can work together to get many homes signed up simultaneously, there may be benefits from the high level of coverage this brings, but this is no substitute for cellular coverage.

Alternatively, as discussed in Chapter 3, users may deploy femtocells in their homes. In this case, there are some advantages to the operator in making access open since this can reduce interference problems and offload data from the cellular network. This is not a true community network because, while the femtocell is deployed by the home owner, it is operated by a cellular operator.

Community networks do not change the overall architectural picture; they just make it easier for users to connect to small cells by spreading the deployment effort over a large number of individuals rather than a small number of operators.

10.5 Implications for the wireless value chain

Consumers tend ultimately to get what they want. While corporate interests may deliver something different for some time, in a competitive market eventually a disruptive player finds some mechanism to change the value chain. For example, Apple have demonstrated how they are able to change the rules of the cellular industry (although not in a manner that fully realises the 'always best connected' concept) and others can be expected to have a similar effect over time. In some cases, even the cellular operators themselves might change if they perceive that their current business model is not delivering the returns that they require (and this is starting to happen already).

A value chain better aligned with the interests of the user would be one where one or more layers of the hierarchical network were owned by wholesale providers. These organisations would maintain the networks as 'bit pipes', providing data-transmission services at a wholesale level to a range of new service providers. The service providers would not own any network resources themselves, but instead would buy a mix of data-transmission services from the bit-pipe providers and bundle these together into an attractive offering for the end user. Others in the value chain might include application developers who made appropriate applications available through 'apps stores', home-network maintainers and so on.

This realignment of the value chain is critical for the future of wireless communications. Without it, inefficient technical solutions will be employed, significantly hampering what can be achieved. When and how it will happen is unknown, since it depends on the actions of key individuals running the major players.

11

The green agenda

11.1 The scale of emissions from wireless systems

Across all areas of human activity there is increasing realisation that the emission of greenhouse gases is causing undesirable global warming effects and coupled with this is a desire, and in some cases a requirement, to reduce emissions. This applies to the wireless sector as much as to any other.

It is not entirely clear exactly what contribution wireless makes to greenhouse-gas emissions. Doing such calculations is always very difficult, in particular in defining what should be within their scope. For example, it is clear that power consumed by base stations and mobile-handset chargers should be included, but what about power consumed by laptops that have embedded wireless devices? Deciding whether to include manufacturing and recycling costs can be problematic, as can attribution of costs in areas such as the paper used by cellular operators in the normal course of their business. All of this means that any attribution will be approximate and may change over time, both as more information becomes available and as decisions on the elements to include change.

There is basic agreement on the broad contribution of information and communication technology (ICT), with the total global carbon footprint in 2008 estimated to be about 800 $MtCO_2e$ (mega-tons of carbon dioxide or equivalent) or approximately 2% of global emissions. This is predicted to grow to about 1,400 $MtCO_2e$ by 2020, or approximately 2.8% of global emissions at that date. ICT includes not just all communications but also all uses of computing devices and hence is a much broader category than wireless.

The contribution to ICT from global telecommunication systems including mobile and fixed communications devices but excluding TVs and TV peripherals is estimated at about 29% or approximately 230 $MtCO_2e$. TVs and related

Table 11.1 *Emissions from key wireless networks [1]*

Communication system	Proportion of total emissions in country	Relative emissions (3G networks = 1)
3G networks	0.03%	1
2G networks	0.1%	3.3
DTT transmisson	0.01%	0.3
Domestic TV equipment	0.57%	19

peripherals contribute of the order of 700 $MtCO_2e$, nearly as much again as the total for global ICT. Assuming that half of the communications figure relates to wireless and adding in TV results in a number very similar to that for ICT (which does not include TVs). Hence, we conclude that the use of wireless contributes of the order of 2% of global emissions. This makes wireless a minor, but not totally insignificant, contributor to global emissions and an area where increased take-up of wireless services might result in an increasing contribution in the future if nothing is done to prevent this.

There are many hundreds of uses of wireless systems and to analyse them all would be a task of enormous magnitude. However, it is clear that emissions are dominated by those areas with widespread consumer equipment, since this results in both emissions during manufacture (and disposal) and energy requirements when in use. For example, we can safely assume that public cellular use results in much greater energy consumption than private business radio of the sort used by taxi companies and many other entities. This is because the handsets are similar but there are probably two orders of magnitude more cellular handsets than business radio handsets.

Table 11.1 shows the emissions from key wireless networks for a typical developed country.

For phone networks the emissions related to the handset, predominantly those for manufacturing, are approximately equal to the emissions from running the network when the lifetime of the mobile and the network is taken into account. This is highly approximate and depends on many factors, but suggests as a rule of thumb that improving the emissions related to the handset is about as important as improving the emissions related to the network.

11.2 The role of wireless in reducing overall emissions

Many have suggested that, while wireless systems undoubtedly consume energy, they may enable savings in other areas that more than offset the emissions they cause. An often quoted example is the use of video conferencing to reduce the need for international travel (although this tends to use wired rather than wireless systems). Other examples are the ability to work remotely, so saving on commuting, the ability to avoid congestion by re-routing around it, the ability to replan itineraries dynamically, thereby avoiding wasted journeys, and so on.

However, for each of these examples there are questions. For example, while home working may result in lower emissions from travel, it may require the home to be heated during the day when it would not otherwise need to be heated, offsetting the savings. Better communications might actually lead to more travel by making the booking process simpler and making the flights less expensive since airlines can become more efficient – budget airlines are based in part on Internet booking and printing of boarding passes. Better communications and congestion avoidance may actually allow more journeys to be fitted into the day, increasing productivity but also increasing emissions.

These counter-examples illustrate the great complexity in predicting how changing certain factors will ultimately influence behaviour and how this will impact on emissions of greenhouse gasses. So, while wireless communications might enable a reduction in emissions in other areas such as transport, this cannot strictly be used to justify greater emissions from wireless systems or to offset existing emissions.

11.3 Making wireless systems greener

Many involved in the wireless industry have major incentives to reduce emissions. Operators can reduce their operating costs if they use less electricity. Manufacturers can differentiate their network equipment if it is 'greener'. Consumers are interested in phones that use less energy and, in particular, in low-power chargers, both to reduce their electricity bills and to reduce their carbon footprint. Governments often have emission targets and may impose restrictions on particular sectors in order to reach these. Some of the key initiatives are described below.

Networks

Reduced amplifier power. One of the largest users of power in base stations is the amplifier. This can often consume more than half of the power of the site and also contributes to the need for active cooling (see below). Hence, reducing the power needed for an amplifier can have one of the largest impacts on total power requirements. Base-station amplifiers are often only about 15% efficient (that is, only 15% of the power they consume is delivered to the antenna; the rest is dissipated as heat), so there would seem to be substantial scope for improvement. However, RF amplifiers typically need to have very good linearity – that is they need to avoid any significant distortion of the signal that they amplify. If their linearity is compromised, this results in unwanted spectrum emissions that manifest themselves as interference and may reduce overall network capacity or cause interference with other spectrum users. Typically, in order to achieve good linearity, amplifiers have to be operated well below their maximum power levels, which results in inefficient operation. The modulation method of OFDM used in LTE can add to the problem, since OFDM has a greater peak-to-average ratio[1] than most other modulation methods, requiring amplifiers to be operated even further below their maximum power levels ('backed off' to a greater degree) in order to avoid interference. The use of MIMO antennas might also result in the need for a separate RF amplifier for each antenna, further increasing energy requirements. This is another reason why MIMO solutions might not be as widely deployed as some are currently indicating.

There are many possible solutions to this problem. For example, amplifier linearisation seeks to use feedback that compares the output with the input and adds a correction signal to the input such that any non-linearities in the output are corrected. Many companies are currently researching this area, since any better solution will be in much demand.

Another approach is to reduce the losses that occur after the RF amplifier such that the RF amplifier does not need to provide such a high output level. Many of the losses occur in the cabling that links the amplifier to the antenna, particularly on tall masts where cabling runs can be long. A solution here is to mount the amplifier at the top of the mast (such amplifiers are known as 'mast-head amplifiers') so that the cabling run is much shorter. This brings disadvantages from a maintenance and installation viewpoint, but these can be overcome by careful design.

Using processes such as this, over time base stations have become steadily more efficient, although many of these gains tend to be lost with the introduction of new generations of technology and have to be relearned.

Removal of cooling requirements. Another large part of the power consumption of base stations is often air conditioning or other similar cooling

[1] To understand this, consider an OFDM signal made up from 128 sub-channels. On average, for each symbol about half of these will be a 0 and half a 1. Since the 1s partly cancel out the 0s in their power output, the overall signal level can be quite low. However, on rare occasions the vast majority will be 1s, requiring a very much greater signal level. The more sub-channels there are, the more extreme the peak-to-average power ratio will be.

systems. This is needed in order to keep equipment temperatures below 40–50 °C, above which components can malfunction. The need for cooling comes from the inefficiency of amplifiers, which generates heat, which then requires more energy to remove. Hence, there is effectively a double gain from reducing amplifier power requirements.

Cooling requirements can be reduced in a number of ways. The most obvious, as already mentioned, is to improve the efficiency of the amplifier and hence reduce the amount of heat it produces. Another is to use components that can tolerate higher temperatures, reducing the amount of cooling which has to be applied. A third is to design equipment racks so that natural cooling can be effective, such as fitting a chimney-like structure that rapidly channels the warm air away from the base station.

Through a mix of all of these different approaches, designers of base stations are gradually removing the need for active cooling, especially for base stations with fewer channels and less capacity.

Reduced obsolescence. As well as the energy consumed during operation, base stations and other network equipment require energy for their manufacture and disposal. The longer the lifetime of the equipment, the longer the time period over which this energy can be spread. Of course, longer equipment lifetime is also in the interests of the operators, who do not wish to replace their equipment frequently. Obsolescence can be reduced with modular equipment such that cards can be swapped out when new technologies are introduced but the bulk of the equipment remains unchanged. However, it remains very difficult to design equipment that can accommodate a complete change of standard such as from 3G to 4G without replacing most of the components.

Switch-off during quiet periods. A simple solution is to shut down all or part of the base station when it is not needed. It might be possible to shut some base stations down entirely during periods of low demand with coverage provided from other nearby base stations, or it might be possible to shut down some of the processors and channel cards.

Consumer equipment

Longer replacement cycles. Much of the energy use on the consumer side comes from the replacement of devices such as handsets and TVs. As with the network side, the longer the time period that can occur before device replacement, the lower the energy usage. Broadly, this is a consumer-education issue, since it is consumers who choose to replace devices, although changing the nature of the subsidy of handsets in some markets might reduce device-replacement cycles.

Energy-efficient chargers. Much in the news has been the energy requirements of chargers, particularly phone chargers. Many, it appears, are left plugged in permanently and phones connected when charging is needed. However, many chargers consume power even when they are not charging a device. Improving chargers to have almost zero power consumption when not charging and to be as efficient as possible when charging can make a difference, albeit a small one.

The move towards common chargers across all mobile devices might reduce the need to supply a new charger with each new phone, reducing the energy required to manufacture chargers. Again, this will make a small difference.

Lower standby currents. Similar to energy-efficient chargers is for devices permanently connected to the mains power to have low power consumption both when in use and when on standby. The biggest gains appear possible on standby since until recently manufacturers paid very little attention to this and often designed energy-inefficient but low-cost chargers.

Integration of functionality. Integrating multiple devices into one box can save power because only one power supply is needed and other savings can often be made. For example, integrating a femtocell and WiFi router into the same box can save power and even more gains can be made if older systems such as DECT cordless phones are removed from the home.

11.4 Implications for future wireless systems

It is an unfortunate truth that, in just the same way as more people cause more emissions, so do more wireless networks. The more networks there are in a country the greater the emissions will be, even if each network introduces mechanisms to lower its emissions. However, a mix of technological progress and governmental desire for competition has resulted in an ever-growing number of wireless networks – for example, in some countries there are eight to ten 2G and 3G networks in existence and the intention to introduce another three to four 4G networks without shutting down any of the existing ones. Add in TV networks, emergency-service networks, radio networks and many others, some of which are also rolling out variants in different technologies, and it is clear that, unless there is a radical change, the number of wireless networks in a country is likely to continue to grow over the coming years. However, if there is sufficient pressure to reduce emissions then it is possible that this trend might be reversed. This section discusses some of the possible approaches that might be adopted and considers the implications for wireless networks.

There are ways in which networks can be consolidated or switched off more quickly. One approach, particularly relevant to broadcasting, is to minimise the length of time during which material is 'simulcast' – that is, the same material being broadcast simultaneously in more than one format. Simulcasting is prevalent at the moment during the switchover from analogue to digital broadcasting, with the two networks operating in parallel for five years or more, allowing consumers a long time-horizon to switch from one technology to another. Simulcasting for extended periods does ease the transition for consumers, but brings penalties in terms of both poor spectrum utilisation and higher emissions, since power is needed for two separate networks. For major switchover programmes it is likely that some period of simulcasting will be needed to ease the consumer adoption of new equipment, but, if there is a growing concern about greenhouse-gas emissions, then in future the time period for simulcasting might be made significantly shorter.

There is a similar issue in cellular networks, where 2G networks are kept turned on despite the fact that 3G is available and at present it seems likely that both 2G and 3G will be left in operation when 4G is introduced. This is not strictly simulcasting, since the same material is not sent across different networks – a phone call will be handled by only one of the network technologies, depending on which is best placed to serve the user. However, it does result in higher emissions and may be less efficient in terms of spectrum use. There are various reasons why operators keep multiple networks running. One is the same as for broadcasters – it can take time for users to switch phones to those capable of working on the new network. However, this is generally much more under the control of the operators, who often subsidise phones and could encourage more rapid migration if they wished by directing their subsidies towards phones capable of operating on the new networks. Generally they do not do this. Another reason is to attract roaming traffic from other countries, which is often particularly lucrative because of the high international call charges that are imposed. Since most phones have as a 'fallback' a 2G capability, running a 2G network helps to increase the chance that roaming phones will find and log onto such networks. Legislation can be a problem in some cases – in much of the European Union until recently the 900-MHz band was mandated for GSM use only, preventing operators using it for 3G. This has now changed, although operators have not been quick to make use of their new flexibility. Another problem can be the difficulty of 'refarming' – that is, changing the use of one piece of spectrum from an older to a newer technology. Newer cellular technologies tend to have ever wider carrier bandwidths – GSM had a bandwidth of 200 kHz, 3G operates at 5 MHz and 4G uses bandwidth up to 20 MHz. This typically means that most of the use of the older technology needs to be turned off before the newer one can be introduced, whereas operators might prefer to turn off a fraction of the older technology and gradually introduce more of the newer technology as subscribers migrate. This can make a switchover a major logistical problem – although not an insurmountable one, since many operators have switched off a first-generation network and migrated their entire subscriber base onto the second generation.

While these are real problems, all of them can be resolved if there is sufficient incentive. Generally, it has been that incentive which has been lacking to date. Operators have not been under any pressure from regulators, governments or users to switch over and the economic case for doing so has been relatively weak, since the newer technologies did not result in significant cost savings compared with the older ones. It might be that pressure to reduce emissions changes this, with the result that the period of running multiple different network technologies is reduced. For example, operators might wait longer before deploying a new technology in order to ensure that it is ready for rapid widespread deployment but then, once it has been deployed, aim to migrate subscribers onto the network within one or two years, enabling them to switch off the old technology.

Another approach to reducing the number of networks is for operators to share them. If multiple operators use the same site then this reduces the emissions due to site construction, such as those resulting from the pouring of the concrete tower base. If operators go further and share the electronic equipment at the base station then this can bring reductions in the overall energy requirements. Moving from four separate networks to two shared networks would not halve emissions, since the base stations would still need to transmit as many carriers and process as much traffic, but it might perhaps reduce emissions to 60%–70% of the amount that would have occurred, depending on whether they have spectrum allocations in the same frequency bands, which tends to enable greater sharing of masts.

Network sharing, at least at the mast level, is already widespread in many countries. This is not in order to reduce emissions but to reduce costs. Sharing masts can remove almost half of the cost of deploying a new site with limited downside. Equipment sharing is less widespread because the additional cost savings are relatively small, while the operational difficulties are much greater. These include the need to adopt the same software load and feature set as a competitor and the difficulty in changing supplier. Many commentators expect widespread network sharing for 4G networks, with perhaps only one or two networks deployed around the country. This is because the profitability of a cellular network is continually falling as call costs drop and operators are less inclined to make massive investments in new networks and technologies for which the returns are becoming increasingly uncertain. A desire to reduce emissions might add extra impetus to this.

Some have postulated a much wider network share with cellular towers used to broadcast TV, radio, emergency-service networks and more. While there is some logic to this approach of minimising the number of masts and reducing the interference between networks through co-siting (hence areas of weak signal from one network do not coincide with areas of strong signal from another), the practical difficulties of achieving this seem much greater than any gains. In some cases mast sites are shared across multiple network types where they are suitable and the commercial conditions are favourable. This will certainly continue and may steadily grow. Expecting a major change such that all non-cellular network sites are shut down seems impractical, at least as a forced migration in a short time period. Only over a longer time might networks move in this direction, if it makes commercial sense.

As mentioned above, changing consumer equipment results in a large increase in emissions. Regarding mobile devices, consumers tend to change their equipment for their own reasons, but with TV and radio receivers some of the change is a result of the broadcasters (or regulators) changing the broadcasting standard – for example, the move from analogue to digital broadcasting will have caused some to replace TVs, PVRs and similar that they might not have done otherwise. This clearly causes a conflict for those managing wireless broadcast technologies. On the one hand, introducing new technology such as MPEG-4 encoding allows more efficient use of the spectrum and the introduction of new services such as HD broadcasting. On the other hand, it encourages consumers to purchase new equipment or replace equipment when they would not otherwise have done so. It also, as mentioned above, implies a period of simulcasting, with the emissions implications that involves. How these conflicting requirements will play out in the future is far from clear. Making no changes to broadcast standards ever again seems unlikely, although it may be that, as viewing moves towards IP-TV, terrestrial wireless TV networks will not experience any more major updates. At present the environmental concerns are generally not considered by those who have responsibility for decisions on new broadcast technology and it may be that this situation continues. If regulators and broadcasters are given environmental objectives as well as their other duties, they will have to decide how best these should be balanced. Perhaps all that can be concluded at the moment is that it will become progressively more difficult to introduce new technologies that require substantial new consumer equipment.

Within cellular networks there are some strategies that help the network operator reduce emissions. One is to assign lower frequency bands, which have better propagation, so that fewer base stations can be deployed, especially in rural areas. (In urban areas networks are normally constrained by capacity rather than coverage, so having lower frequencies does not provide any advantage.) This might bias regulators towards considering assignment policies not just on an economic basis but also on an environmental one. However, there is limited room for manoeuvre here – cellular operators would prefer lower frequencies in any case because their use reduces the costs of network roll-out, so there are strong market and commercial incentives already to seek to clear appropriate spectrum. It is hard to see environmental duties making any significant difference here.

Another strategy is to make widespread use of smaller cells. For example, studies for Ofcom [1] have shown that, if an operator wishes to achieve indoor coverage, they can do so much more energy-efficiently through the deployment of femtocells within the home than by increasing the density and power levels of the outdoor macrocellular network. This seems intuitive – transmitting power exactly where needed and in the process avoiding the losses associated with signals passing through external walls will clearly result in lower transmit-power requirements. However, as noted earlier in Chapter 3, there are still some outstanding questions around femtocells. The same gains could also be achieved using WiFi within the home and, indeed, within other buildings as well, a path some operators are considering for economic reasons.

Summarising the potential impact of environmental concerns, it is clear that at present they are a relatively weak force compared with other considerations such as economics and consumer choice. For cellular networks many of the changes that would reduce emissions align with the economic interests of the operators and other players, and so would tend to happen in any case. For the broadcasters there is some conflict in the introduction of new technologies and it remains to be seen whether this will have much impact. For the manufacturers the reduction of power in their network equipment, especially base stations, will continue to be of importance, and research into better RF amplifiers can be expected to be a key topic for many years.

Within the industry at the moment the 'green agenda' is often given great prominence, with suppliers and operators striving to demonstrate their green credentials. From the currently available corporate literature it might be deduced that environmental issues were one of the greatest challenges (and opportunities) facing the wireless industry. The analysis above suggests that this is somewhat overstated and that companies may just be responding to the perceived political and public mood. The use of wireless in the future may well result in lower levels of emissions than today, but this will be achieved without any fundamental changes in networks or regulation.

Reference

[1] http://www.ofcom.org.uk/research/technology/research/sectorstudies/environment/environ.pdf.

PART II SOLUTIONS

12

The future of users

BY RICHARD HARPER[†]

12.1 Introduction

It might be assumed that a book on wireless futures would concern itself solely with technological matters, while users of that technology are left for a different text. At this point in the book, however, we turn to those technologies regarding which the role and changing behaviours of the user are central. A concern with users helps define what those technologies might be, how they will evolve, and what changes they might bring about in user behaviour that will in turn have implications for the technology. In this chapter, we present an overview on the importance of user behaviour in this regard before offering some high-level prognoses concerning the future. Subsequent chapters will deal with particular technologies and themes, such as location, health and transport.

As will become clear, one cannot separate the evolution of user behaviours entirely from the possibilities that the wireless landscape affords – the two are inseparable. Nonetheless, an emphasis on the user can highlight issues that are sometimes neglected in wireless research. This can help guide insights into the future, which is our task here. The interface between devices and services and the user will be central to this, but so too will be the changing trajectories of actions enabled by new hardware and services. The interface is merely the prism for both what users can do and what they want to do, both of which broaden through time.

[†] Richard Harper <mailto:r.harper@microsoft.com> is a Principal Researcher at Microsoft Research in Cambridge and co-manages the Socio-Digital Systems <http://research.microsoft.com/en-us/groups/sds/default.aspx> group. He is the author of ten books, including *Texture: Human Expression in the Age of Communication Overload.*

12.2 Historical background: the first users

It might be said that until quite recently users have been a taken-for-granted animal in the wireless industry, known to be the ultimate paymaster but not a creature that has to be investigated in any great depth. That this is so might not be surprising – the mobile user is after all simply a person, and therefore only good sense is required to understand why they might want to communicate or, say, spend their money on wireless technologies and services. However, if one looks back at the evolution of wireless devices and services, and at the changing use those technologies have been put to, it becomes clear that the industry has had more than a passing notion of who the user of its technology might be, one that does not assume that the user it deals with is the same as, say, the one conjured up by the PC industry. This vision of whom the user is, what they want to do, and how their habits have evolved is, in important respects, unique to the mobile domain. This has been reflected in technological development. This has in turn shaped users themselves as they have learnt what the technology can do for them. As they have discovered this, they have, in turn, shaped the technology in response to their demands. The user is not one that has remained static over the years, and neither is the relationship between the user and technology one way.

That there is a relationship between the user as conceived of by the wireless industry and the actual shaping of wireless technology can be seen by reflecting on the model that the mobile industry originally used when it first set up the wireless networks. When, for example, the GSM networks were first planned and developed in Europe, the user was expected to be a business person, and the high cost of the mobile-phone calls would be reflected in the value of the calls these persons would make. High costs equalled high value, in this view. A related consequence of this model was the expectation that usage levels of wireless communication would be low, with calls being infrequent even if they were mostly of high value. Another was that geographic coverage was more important than density of coverage (say in urban areas), since calls would always be important, wherever they were made from. Hence the early networks were designed to allow the travelling business person to make calls from any place, any time. They were not designed to allow huge numbers of callers to contact each other within the confines of, say, cities.

The logic of this also meant that, if users had important business to communicate, then it would also be the case that technologically richer modes of communication would be more desirable. This would allow them to communicate more effectively. Hence, if early GSM networks could offer talk (and text), then the vision of the user that underlay the wireless industry's plans led them to invest in enhanced networks that would supersede the early networks with ones that offered video and video conferencing (together with other services, of course). The costs of such data-demanding connections would be justified by their importance; the efficiencies of being able to see those being communicated with would make the calls more effective for those involved.[1]

12.3 Take-up beyond the business community

Actual use and take-up of the early wireless networks made it clear that this model of the user was too limited. It wasn't the business users who took up the services on offer so much as ordinary consumers, indeed in huge numbers. Also, some years after the implementation of video capacity, there was no shift towards media-rich communications even among the business users of mobile systems. Voice and text remained the backbone of usage as well as the primary source of revenue; video connectivity was an incidental.

The fact that wireless networks were more widely taken up than expected forced the wireless industry to rethink its model of the user. As remarked above, the industry had assumed that communication over mobile networks would have to have some functional value to justify the expense. Though it was difficult to define what functional (or important, or necessary) communication might be, that there was some threshold that distinguished between important and other, less important communication was a starting point for much of what the wireless industry did. The economic models used to calculate the bidding values for the first spectrum auctions were based on these models, for example. Similar models were also used to calculate the capacity of these networks. The unexpectedly large use of mobile phones for apparently non-functional uses forced a revision of this threshold upwards. The industry's response was to raise the proportion of communications which could be treated as functional. This in effect reversed the order of the previous assumptions that had underlain the industry's view of the user – instead of a definition of the functional communication that users sought to undertake being the determinant of mobile-communications volumes, it was the mobile traffic that determined the functional communication of users.

[1] Interestingly, this idea had its echoes in the view developed at the same time within management science called media-richness theory. This holds that the more important the act of communication, the greater the desire on the part of those involved to use richer communicative formats. More important communications deserve richer technological affordances, in other words.

12.4 The semantics of user behaviour

This may be little more than an interesting historical aside, but it also led the wireless operators to recast their approach to understanding the user. This had numerous consequences. One was recognition that communication might have values other than the strictly functional. Early research on texting had led the operators to think that it was cost that motivated take-up, for example. But later research showed that this was only one factor. Another had to do with the gift-like qualities of text, and how the receipt of a text could necessitate a text in return. If this was the case with text, then operators began to think that this might also apply to other communications too. Voice calls could lead to more voice calls, for example. In other words, more communication could make more communication.

This led the wireless industry to invest in ways that would increase the likelihood that a person would communicate, rather than emphasising the possibility that, with mobile networks, important communications could be made any time, any place. Consumers were not sold mobiles so as to contact the emergency services alone, but to be in touch more generally. Degrees of 'propensity to communicate' became important metrics used in marketing campaigns and service design.

Relatedly, the role of technology shifted from being intended to support people when they were mobile, to supporting them wherever they were – even at home or at work where cheaper fixed-landline alternatives were available. Mobile phones became *personal* phones. As a consequence of this, the mobile operators and manufacturers started to emphasise the contrast between traditional place-based communications and their own person-based offerings. The equation of wireless technology with 'personal needs' naturally led to a further emphasis on personalisation of the devices and services offered.

Having moved beyond the original business user, the wireless industry created problems for itself, however. Whereas before they could tailor both the services and the devices to the particular type of user – the well-appointed travelling business person – the situation they had created, one in which great masses of people delighted in mobile communications, required them to satisfy a range of preferences for function, design and style, as well as cost. Given that the market place for networks and terminals was crowded with suppliers and manufacturers, this pushed down the costs and margins inexorably, leading many companies to leave the field. Though the industry had succeeded in broadening the appeal of its products to a wider range of users, this had the paradoxical effect of making business harder to do.

12.5 Early attempts at a wireless web

The shift towards the perception that mobile phones were a personal technology, while increasing the diversity of devices, meant that, though more people had wireless devices than ever, a 'mobile world wide web' did not emerge. The mobile phone became a personal device, but at this time it did not become a personal computer that linked the users to the web.

Early attempts to provide usable standards for information exchange in the later nineties, such as WAP, didn't alter this. Most, like WAP itself, were badly implemented and in any case reflected misunderstandings about what users might want to do with their mobiles and the information that could be accessed by, or delivered to, them. Most of these efforts assumed that mobiles would become, with the right standards and GUIs, simply windows onto the web. In this view, services and products on the web needed to be resized for the screen and interaction limits of a handheld device. It was only with the advent of the iPhone that users starting accessing information and non-messaging services in large volumes, and when they did so their behaviours had rather distinct characteristics. We shall come to these shortly.

12.6 Beyond talk

Whilst the take-up of information-based services remained disappointing for the wireless industry, similar disappointment arose with regard to mobile video telephony. Though video telephony had been available at high cost over fixed lines for some time, and though the take-up of these systems had not been as high as expected, the wireless industry was nevertheless convinced that its offerings would lead to a substantial increase in the use of video as a mode of communication. It was assumed that video would appeal, given the right cost and ease of use, since being able to see would naturally bring those involved 'closer together'.

Though videocall traffic did emerge (and indeed it continues to show small year-on-year increases), it nevertheless remained much less than expected. Even today, in 2010, no more than 2% of traffic (according to some estimates) is now given over to video connections between callers.

One reason for this had to do with the fact that video calls are perceived as still expensive, and the benefits of them, though recognised, are not equal to the added cost over voice. Another had to do with lingering standards and network-interoperability issues that could make video connections difficult to achieve.

Beyond this, however, other factors begin to show themselves, which suggested that there were more profound limitations with video connection. These were of an essentially psychological and sociological kind. For example, research suggested that video connections broke various rules of public and private behaviour. A video call to one person might be acceptable to that person, but others, who could be seen by the caller, might find the looking of the remote stranger (through the mobile) invasive. Some researchers suggested that this problem would abate. In their view, this aversion was similar to the complaints made about the noise early mobile phones created in public space. Just as people got used to the sound of mobile phones ringing and, thereafter, the mumbling of people talking to remote strangers, so people would get used to video calls invading the visible space of public settings. But this change does not seem to have occurred.

It could be, therefore, that there are more profound limitations to video telephony on wireless networks. One obvious ground for these limits has to do with the ergonomics of wireless devices. Webcams are fixed in particular places that allow participants in a video connection to place themselves so as to be seen correctly. Mobile devices held by the hand inevitably move, blurring images. The devices have to be held away from the head too (to get sufficient viewing angle), and this in turn requires that the speaker volume needs turning up. This makes video calls intrusive for reasons unconnected to sight. It is the sound that can be the source of user resistance, not vision.

Besides, it has also been suggested that the nature of the communications made over the wireless networks, particularly through mobile phones, is such that the added value of a video connection is not particularly beneficial. Evidence for this claim derives from the fact that video communications do seem to be successful in workplaces. Here, the richness afforded by video would seem to satisfy requirements that mobile video does not. One reason for this, perhaps the most obvious one, is that work video conferences often have important matters to address. Users may feel that the richness of video can enable better decisions to be made. In contrast, mobile phones are used for more lightweight, personal and micromanagement-type purposes. Here, video connection does not add any great advantage. At work it is also the case that video-conferencing facilities are combined with comprehensive ancillary equipment – document cameras, stereo speakers and so on.

A further reason why video is more acceptable at work is because concerns about invading the privacy of a person standing nearby are much less – if relevant at all. All those who attend the meeting will know beforehand that a video connection will be made; often rooms are allocated to video-conferencing activities, so there are not any non-participants in sight.

In short, wireless video connectivity might not flourish because of the content of mobile communications and the contexts of use. Mobiles are towards the personal end of communication, video conferencing at the more formal and organisational.

But just as this is the case, then it might also be the case that extremely personal uses of the wireless systems for video connectivity would be on the up. Intimate communications can be facilitated. There is no evidence that this is happening to any large extent, though. Much more evidence is available suggesting that wireless devices are being used to capture personal video. This is then shared or posted on social networking sites. Indeed, the volume of this form of video exchange is increasing rapidly. From the industry's point of view this raises a number of concerns. This is not duplex video telephony and hence not subject to premium charging. Such videos, though mostly taken with mobile phones, are often delivered over WiFi and/or fixed-line access to the Web to be viewed and exchanged. Often these video files are then accessed by a mobile device, but again through WiFi facilities. In this scenario, the wireless networks are entirely outside the loop.

Despite this evidence, the wireless industry is still convinced that video telephony, undertaken through and via mobile devices and networks, will flourish eventually. It is certainly true that video is being used increasingly, but the character of the usage patterns suggests that it will not be of any advantage to the wireless industry, and, indeed, there is concern that it might distract the industry from developing services and opportunities more suited to the wireless setting.

12.7 Moving beyond person-to-person communication

Meanwhile, by the middle years of the first decade of the new century, various new types of smart phones and communications-enabled PDAs had appeared on the market. Each of these sought to enable and uncover the kinds of services and applications that would lead users away from person-to-person communication towards more information-centric behaviours. The industry still believed that these behaviours would be analogous to those activities users undertake on the web via PCs. New standards were sought in operating systems, in web-server architectures and in programming languages that would learn from the mistakes made when WAP was developed. All of these efforts would lead to the emergence of a mobile-centric ecology. Many of the major terminal manufacturers developed devices with just this in mind, as did the major operators who developed their own mobile-focussed portals for services and products. As before, success was not as great as envisaged. Systems-integration issues appeared central to this. Cost sensitivity on the part of users was also a concern, combined with what was thought to be users' conservative attitude towards new services and products.

However, two complementary changes affected this situation to the benefit of the wireless industry. One was the introduction of a device from a manufacturer that had not played a major part in the wireless industry before – or at least not since the early nineties when it had launched its Newton PDA – namely Apple, and its iPhone. The other was the emergence of social-networking applications that were device-independent, just as they are independent of the wireless industry. Just like the iPhone, social networking altered the landscape of human communication. This has implications also for the future of the user.

The success of the iPhone was at first dismissed by the wireless industry as marketing hype, though many admitted it did offer improvements in levels of usability. But the real significance of the iPhone lay partly in its design and functioning and partly – perhaps more importantly – in how it led to the creation of a vibrant mobile-application ecology, one that had simply not materialised before. One of the reasons why the iPhone was little regarded by the rest of the wireless industry when it was launched was because of the limits of its OS. This supports only one application at a time. However, just as the selection of the very simple Epoch operating system as the basis of the Symbian OS had the advantage of forcing economy of design on the part of software engineers, so the limits of the iPhone OS forced greater emphasis on elegant application design and integration. It was this, among other factors, which helped the iPhone to offer substantially better levels of usability than had been the norm before. This was combined with powerful graphics processing. The overall effect was a device that immediately appealed to users, who felt the designers had at last made handheld interaction intuitive, and with graphics that charmed users with vibrant colour contrasts.

Nevertheless, the appeal of the iPhone would not have led to the flourishing ecology around it if Apple had not allowed third parties to develop and offer applications for it through the Apple Store. At first Apple was very reluctant to do this, thinking (like most players in the wireless industry) that success could be guaranteed only if it maintained control over the applications available for it. Apple thought it would be better able to identify, design and implement applications than third parties. Yet it was really only when Apple allowed third parties to take up this burden (admittedly with Apple's own development tools – and hence analogously to the use of Microsoft tools for third-party developments on PCs) that a remarkable new ecology emerged.

12.8 Diversity of use

The emergence of this ecology does not corroborate the expectations that previously underlay models of the mobile user. For this new ecology resulted in the production and purchasing both of applications that fitted the previous models (offering well-understood location-based services for example, ones that lead the user to the nearest restaurant and suchlike) and of others that had hitherto simply not been envisaged. Indeed, the success of many applications available via the Apple Store defied any normal marketing logic. These include screen-saving applications that demanded that the user interact with objects to keep the screen saver 'alive', for example, and music-making applications that used intelligent algorithms to interpret data from movement sensors in the device. These allowed users to create what can best be described as playful musical dialogues, but which had little musical merit.

The upshot of this was that this new ecology produced or encouraged consumer behaviour that was and continues to this day to be enormously diverse. Users learnt to express themselves in myriad ways. No one iPhone user had the same set of applications installed as any other. The world enabled by these various ecologies was one made common by its distinctions.

Whatever this says about behaviour, this was nevertheless good news for the wireless industry, illustrating that there is a substantial business beyond voice and text. As importantly, it also illustrated that this world is not a strict analogue of the web, with many of the successful services and applications on iPhones and other smartphones being bound to the particular affordances of the handheld devices in question, and to the social contexts of connectivity they enable.

12.9 Social networking

At the same time as this was happening, an unrelated development affected the communication habits of people. If, through the nineties and the early part of the new century, the wireless infrastructure was used to deepen and cement the bonds between people which already existed, then the emergence of social networking led to new digital identities that had hitherto not existed or even been imagined. These identities were more than mirrors of real selves, since they provided new opportunities for users to continually construct how they were seen in the digital world (through regular postings and alterations to their profile), where these efforts would in turn foster comment and communication with others, also through the digital world. Social-networking sites are not places where people simply represent themselves, but rather came to be places where the daily management of digital identity fostered new conversations and communications. Digital sociality produced digital connections.

12.10 Prognosis

In overview, the historical evolution of the wireless user could lead to complacency. Even though the industry may have been faltering at some points, user behaviours have in the end resulted in an ever-increasing demand for wireless services, with voice and message volumes up and the use of data up prodigiously too. Blips in understanding have not altered the basic soundness of the wireless industry's view. But this should not lead to neglect of what the evolution of user behaviours might be. The past 10 or 20 years can provide insights and suggestions as to how this might evolve.

For example, if it is the case that an increasingly larger proportion of time given to being in touch is via social-networking portals, then it might be that the providers of those portals will come to gain a greater control over the messaging formats. This might affect the wireless industry.

As a case in point, Facebook users can exchange messages with those they are linked to either by letting the service send an email or SMS, or by keeping those messages entirely within the Facebook application. Many operators now provide Facebook access on the mobile devices they sell. Users can thus access and communicate to their friends either via a third-party channel, such as SMS or email, or through Facebook itself. If they choose the latter, then the mobile industry is being cut out. Though the costs of messaging through the social network might be covered by network charges, the point is that the wireless industry will be losing its role as a messaging channel and hence its ability to charge premium rates.

It may also be that the use of social networking will have a changing profile. Currently it looks as if social networking increases the volume of messages produced between members of a network. But it might be that this is a temporary phenomenon, with the excitement of having digital identities and commenting on them merely a reflection of the novelty. Once these become the norm, messaging might decline and even reach a quite low volume. This, too, will be a concern for the wireless industry, which might misjudge the growth of messaging that social networking will create.

News and other media content might also be affected by the widespread use of social networks. Traditional newspapers create a social network of sorts, a network of like-minded people concerned to keep up with the world's affairs. Social-networking services might come to offer news content to their users, broken up into various groups or affiliations, audiences of the like-minded who are connected via 'friending' (i.e. are connected via the social network and their new grammars and vocabularies of friendship). Not only will this have implications for the future of traditional news media, it might also be a revenue opportunity for the industry. It might be third parties doing business where the wireless industry serves as merely the agent, or it could be the industry itself. To date, however, the wireless industry has not shown itself adept at adopting these new roles and it has tended to confine itself to being the vehicle of messaging, not a generator of message content – as in this case, news-type content.

The industry might also need to be wary of thinking that there will be no end to the delights that digitally mediated forms of expression provide. After all, the scale of social networking, combined with other forms of communication, has led some to claim that a threshold will be reached beyond which communication cannot go. In this view, there will be no more slack in human affairs to allow further communication. The wireless industry might need to answer and deal with such concerns. A failure to do so might leave the way open for regulatory restraints or even effects on consumer behaviour such that the appeal of being in touch gets replaced by the desirability of being *out of touch*. Of course, like the concerns about the medical impact of wireless connectivity, the actual impact might be negligible – but the industry cannot assume that this will be so.

In short, connectivity between people made through social-networking services might have other somewhat surprising consequences for the wireless industry. We will deal with them in more detail in Chapter 14.

Other consequences may have less to do with the impact of social networking or smartphones, and be more a reflection of how the wireless industry has, over the past 20 years or so, altered the landscape in ways that alter the relative value of things – inverting what might have been important 20 years ago. If it is the case that the wireless systems eventually produced a setting in which users treated the technology as personal, then another consequence of the changes that the wireless industry has brought about in user behaviours is in users' attitude to location. After all, if a user can communicate wherever they are, then where they are must diminish in significance. However, once the wireless infrastructure has been widely adopted, the value of place-centric services and experiences might alter and give services bound to place a different importance.

Until very recently, the wireless industry stressed the possibility of working anywhere, anytime. This is certainly something that many users now take advantage of. Wireless devices and services allow seamless access to work email and document repositories. However, although the number of those who do work away from the office may be somewhat larger than it was, say, 20 years ago, it is also the case that organisations are beginning to recognise the importance of people working together. In a world in which organisational communications, documents and spreadsheets can be accessed anywhere, the choice of a member of staff to come to work nevertheless shows a moral commitment to the organisation and to their work colleagues. Working together in the next 10 years might not only mean working anywhere; it might also mean being together as much as possible. That this is so might mean that the wireless industry needs to adjust its marketing and product placement. Though the wireless industry has spent a great deal of effort convincing people that they can work anywhere, in the future they might want to emphasise the particular and different values that mobile systems might afford. If businesses and industry have now come to realise that sharing sweat and laughter in a common work space makes for better work, then what do wireless systems do? Do they allow people to share in that sense of being together, or create different connections, beyond and separate from work? Is the future mobile phone even more personal?

The shifted role of wireless technology and the importance of place also point towards the shifting relationship between wireless devices and the larger ecology of devices and technologies of which they are a part. Just as it is the case that the place of work might become more important in the future, then so too it might be that being together at social events will find renewed importance. This might have an indirect impact on mobile services. For example, a great deal of effort has been put into developing video services for broadcasting content over the mobile networks. Here the vision is that people may watch their preferred TV shows anywhere. But it may be the case that people want to watch certain events socially. Though they might want to download and store a video file for viewing at their convenience, they might prefer to watch it with friends and family. In this view, the handheld device might simply be a key to accessing and storing content, while other devices such as large screens in homes might be the preferred devices for actual consumption.

This example also attests to the fact that it is becoming increasingly clear that users do not view their mobile devices as surrogates or proxies for their other technologies. iPhone and smartphone users do not abandon their laptops or PCs (though they might be less likely to buy a new one), but they come to do different things with them.

For instance, the mobile is the preferred medium for sociality, but the mobile is also the medium users have at hand when they are travelling to and from places. The mobile provides a means for filling up this otherwise lost time, but the important point is that users do not fill up this time with the things that they want to do in the places they are travelling to and from. If they could do those things on the mobile then they would not be travelling to them. So, the mobile must let them do other things. One of the great lessons for the success of the latest smartphones (and the growing ecology of services and applications that can be downloaded to them) is how much time people have for indulging in whimsy, in seeking fleeting distraction and in solitary play. Though it might be the case that they justify the purchases of wireless technologies as devices with which to keep in touch, and to do work with when they cannot get to their place of work, it is using the wireless device in between these times that provides the area where the killer applications of the future might apply. If the success of the Apple Store points to anything, it is not that any individual application has altered the landscape of the wireless world, it is that, taken together, sufficiently many of these applications have created a tipping point such that owning an iPhone can ensure that the user always has something to do. It is distraction that is the killer application.

This might seem an unedifying prospect for a technology industry that has spent vast amounts of money and deployed enormous intellectual effort to produce a situation in which people can get on with the practical aspects of their lives. Although the wireless industry likes to think it has been the driver behind the tremendous growth of human communication, there can be no doubt that the user has been too, and that many of the things the user actually wants are mundane, playful, even pointless, one might say.

13

Sensors

13.1 Introduction to sensor networks

Sensors have been widely used for many decades. Types of sensors range from thermostats in homes, to control central-heating systems, through to sensors in cars to warn of a lack of engine oil. Over time there has been a tendency towards an ever-increasing amount of sensors as we seek greater control and convenience. For example, many cars now include sensors to check tyre pressure, windscreen-washer fluid, temperature at multiple points within the cabin, the presence of a passenger of sufficient weight to make the airbag active and so on. In urban areas there is a growing number of closed-circuit television (CCTV) cameras, sensors in buildings to detect occupants and sensors on mass-transit systems to identify users for payment purposes.

Some envisage a world in which sensors are much more widespread. They predict the deployment of additional sensors around cities to measure pollution levels, traffic congestion, temperature and more, or sensors scattered around fields of crops providing information on local growing conditions and allowing precisely tailored application of fertilizers, irrigation, etc. Sensors around the home might detect temperature, light and movement to precisely control the home environment. Sensors in industrial buildings might monitor every aspect of the production process, ensuring a higher-quality output and reduced wastage.

Sensors systems to date have mostly been wired – that is, there is a wire from the sensor to the control system. Most home thermostats are wired into the heating system and most car sensors are connected to the car's wiring loop. But wiring is expensive and time-consuming, so, if sensors are to become much more ubiquitous, they will need to be connected via wireless. Indeed, the goal of many sensor designers is that their sensors can be 'scattered' around and that they will self-connect and operate without any further attention. If these visions of sensors were realised then there could be many tens, hundreds or even thousands of sensors for every person, with the result that there would be many more wirelessly connected machines than wirelessly connected people. It would be a world in which machine-to-machine communications predominated, at least in terms of number of links, if not data volumes.

Before a world of ubiquitous sensors can be realised, a number of challenges will need to be overcome. These include the following.

- Powering the sensors so that they can operate autonomously for many years.
- Deploying a wireless solution that enables them to transmit data cheaply and at low power levels.
- Miniaturising sensors and reducing their cost.

Each of these will be examined in subsequent sections.

13.2 Enabling technologies

The biggest problem for most sensor networks is power. Sensors might remain in place for decades (particularly in the home or a built environment) and should ideally be 'fit and forget'. Periodically replacing sensor batteries would be inconvenient and expensive, especially if this had to be done weekly or monthly.

The actual power requirements depend on an array of factors, including the power needed for the sensor element itself (e.g. the part that measures pressure), the logic circuitry and the power needed to transmit the data, which depends on the amount of data to be transmitted, the range required and the ability of the device to enter 'sleep' modes. Power requirements can cover a very wide range depending on the design choices made.

The options for power are either to provide it within the device, typically via a battery or fuel cell, or for the device to 'scavenge' power from its environment. A range of scavenging options has been proposed, including solar power, vibration and heat differentials. A more detailed description of sensors and their powering requirements can be found in a report commissioned for Ofcom [1]. In essence, the following conclusions were drawn.

- If large batteries could be used (for example, as big as those in a laptop) then a battery life of 10 years was possible.
- With small button-type batteries (as used in watches) the maximum lifetime was about five years, but achieving this was very challenging, requiring ultra-low-power design.

- Energy scavenging can work in some very specific locations. For example, vibration can be used as an energy source for sensors mounted on large industrial machines. However, in general, scavenging sufficient power is difficult to achieve.

These factors are unlikely to change much over time. Battery technology, as discussed earlier, improves only slowly and breakthroughs are not expected. The factors that make energy scavenging difficult are unlikely to change, and low-power design has approached fundamental limits such as the voltage needed to make a semiconductor device switch. This is a serious problem for sensor networks and one that may limit their deployment to those situations in which periodically changing a battery is worthwhile given the other benefits that the sensors provide.

The next problem for sensors is wireless transmission. While sensors could make use of existing wireless networks such as the cellular packet data systems, this would result in rapid battery drain – even making use of all the power-saving options available, batteries would last only for weeks at best. It might also be expensive and would require management to allocate a SIM card and number to each node and then track and bill for its usage. Such a solution is viable only for remote sensors that have large power sources and relay valuable information.

If public wireless networks are unsuitable then sensor systems need to make use of dedicated wireless networks, optimised for their particular application. These need to enable ultra-low-power operation of the sensor and must be of low cost and simple to deploy – essentially 'fit and forget' in the same manner as that in which the sensors would ideally be deployed. The characteristics of most sensors that the wireless network can exploit to achieve these objectives are that

- most sensors transmit very infrequently (hourly, daily, or in some cases only when a problem occurs, which may be less than once a year);
- when they do transmit, they typically send a very small amount of data, often only a few bytes;
- for many applications, the data is relatively delay-tolerant – for example, for temperature readings one might tolerate a delay of minutes (unless they indicated a critical event requiring immediate attention).

A number of standards groups looked at this problem and devised standards to address this particular need. The ZigBee standard was designed specifically for sensors and has very-low-power modes. Other standards, such as Bluetooth, have added low-power capabilities. Predominantly, these allow devices to enter 'deep-sleep' modes in which they do not need to monitor or respond to wireless messages for considerable time periods. This reduces battery drain, although it does make it more difficult for devices to synchronise with the network once they are turned on, which can slow data transmission.

A key question is the architecture of the wireless system. To date most sensor systems have been of 'star' type, with a (powered) central node that collects data from sensors ranged around it. This can work well in a building where each of the sensors might be in range of the central node and is simple and reliable. The alternative is a mesh solution whereby sensors forward information from other sensors until it reaches the central control point. This allows sensors to be deployed over a much greater distance without needing a larger range (which would increase power consumption), but is more complex and less reliable.

We discussed mesh wireless systems in Chapter 5, where many issues and problems were identified. However, the characteristics of mesh systems can work well for sensor networks because the sensors are generally static, data volumes are low and delays can often be tolerated. Careful design is needed to ensure that power consumption does not become high, since sensors would need to monitor transmissions from other sensors in the vicinity in case they need to be forwarded. This can significantly reduce the ability of devices to enter sleep modes. Mesh systems can also be less tolerant with respect to one sensor failing if this is acting as a key intermediate node within the mesh. However, until the issue of powering sensors is adequately resolved, the difficulties with mesh networking may remain somewhat irrelevant.

The final issue with mesh systems is cost-reduction of the sensors themselves. Ideally, the sensor units need to be so inexpensive that they can be liberally 'scattered' around and losing a sensor is not problematic. For many sensors, the constituent parts are not expensive – many sensors can be implemented directly onto silicon using MEMS technology. However, low cost typically comes only with high volume and this tends to be best delivered through standards. Fortunately, a standard has been developed – the IEEE1451 standard for smart sensors. A smart sensor contains its own datasheet parameters in memory and has a standard interface for wired or wireless connections, such as ZigBee or WiFi. This allows sensor modules to be easily coupled to wireless modules, making economies of scale more likely.

13.3 Applications

There appears to be something of a disjunction between those who forecast a world in which sensor networks are ubiquitous and monitor all aspects of our lives and the current situation in which few wireless sensor networks have been deployed, typically only in specific industrial situations. This can be attributed to the difficulties in implementing cost-effective sensor networks, as discussed above, and the lack of sufficiently compelling applications to make sensor networks worth deploying despite their limitations. Applications for sensor networks can broadly be divided into existing sensor markets, where wireless sensors might enable large or lower-cost deployments, and revolutionary applications for which the availability of low-cost wireless sensors would enable a completely new usage.

Some of the major existing applications are as follows.

- Industrial applications, where sensors monitor machinery or production equipment. Sensors add value by ensuring that the machinery is operating correctly. Sensors can generally be wired in these environments, but using wireless sensors would add flexibility, for example when production lines are modified.
- Building control, where sensors can relay information enabling more energy-efficient usage of heating, air conditioning and lighting as well as alerting systems to emergencies such as fires. For new buildings, such sensors can be wired when they are built, but wiring is more difficult for older buildings. Wired sensors can be less flexible if the interior layout of a building is changed.
- Infrastructure monitoring, where sensors are used to check for the integrity of structures such as bridges. These typically need to use wireless because there is limited wired infrastructure, but can have large power supplies.
- Logistics, where sensors can track assets, for example monitoring the movement of packages. Here many sensors can be wired, for example in airports or distribution depots, but using wireless increases flexibility.

Some of the novel applications that have been suggested include the following.

- 'Smart cities' that can monitor the movements of people and vehicles within the city, using this information to assist activities like parking and emergency-service support. Such a network could also be used for automatic meter reading and potentially many other applications yet to be devised.
- Healthcare applications enabling monitoring of people at all times, using information generated by body-area networks and sensors that they might wear (or that might be integrated into clothing).

While the novel applications sound like they might add value, it is very difficult to build a business case for such nebulous concepts, especially when sensor costs remain relatively high. Hence, it seems likely that the main deployment will be a steady advance of the use of wireless sensors in existing applications where their additional cost can be justified.

13.4 Prognosis

Many visions of the future imagine that there will be billions of sensors generating machine-to-machine traffic, which is then transmitted over mesh-based wireless networks (which are often assumed to be using cognitive access, just to add to the challenge!). While applications that could make good use of the data provided by such sensors can be envisaged, there are serious obstacles to their implementation. Powering the sensors looks like an almost intractably hard problem, while using wireless mesh networking itself is challenging. None of the novel applications appear sufficiently compelling to justify the cost and investment needed to overcome these problems.

However, the increasing use of sensors in the home, built environment and industrial settings appears likely. Sensors already add value and, as the cost of energy and the need to reduce carbon emissions grow, the value of sensors in more accurately controlling these environments increases. These will be a mix of wired and wireless sensors. Those using wireless will probably adopt standards such as ZigBee with relatively conventional star architectures. The world of sensors will see steady growth but not an explosion.

Reference

[1] http://www.ofcom.org.uk/research/technology/research/emer_tech/sensors/.

14

Social communications

WITH RICHARD HARPER[†]

14.1 The human habit

It is a commonplace to say that humans are social animals. It is altogether another thing to leverage this fact of human nature to devise new services and technologies to support it. Indeed, the past 10 years or so have made it clear that there is both a need for, and many new business opportunities made possible by, technologically enabled social communications that had not been expected. These have generated and will continue to generate traffic for fixed and wireless infrastructures.

Social connections may be distinguished from person-to-person communications by the fact that they typically entail the broadcasting of messages or, for example, the multiple viewing of single messages on the home page of an individual's social network. Websites such as YouTube, MySpace and Twitter are incredibly popular because they satisfy a basic desire to stay in touch and to know what friends, colleagues and family are up to. It is about being part of a group, not about individual relationships. But satisfying this desire has also created new distinctions in types of social connection and this in turn has cultivated new needs. It is very likely that these will continue to evolve over the next decade or so in ways that will have consequences for technology.

Several forms of digitally enabled social connection can be demarcated. One relates to the sustaining and invigorating of existing social relationships. Here

[†] Richard Harper <mailto:r.harper@microsoft.com> is a Principal Researcher at Microsoft Research in Cambridge and co-manages the Socio-Digital Systems <http://research.microsoft.com/en-us/groups/sds/default.aspx> group. He is the author of ten books, including *Texture: Human Expression in the Age of Communication Overload.*

websites like Facebook come to mind. A second has to do with the use of the web to create new forms of social connection where the parties involved might be unfamiliar to one another. Here blogging and Twitter come to mind. A third has to do with the creation of social connections enabled through web-based services and experiences, such as on-line gaming. Each of these will continue to evolve and this will have implications for technology.

14.2 New directions in social communications

Social-networking sites, the first of the categories above, have been the subject of a great deal of media attention, if only because they are sometimes thought to be mechanisms through which personal safety and issues of privacy are brought into doubt. The use of social-networking sites for grooming of victims of child abuse is well known. But, these fears aside, the popularity of these sites and services has led to suggestions that they might be a major driver of network traffic growth in the future. Though initial evidence suggested that the scale and range of connection made by such sites are novel and had no reflection in any prior form of social network, current evidence shows that these sites are primarily used to sustain and invigorate already existing social connections. Friendships based on geographic proximity, shared schooling and employment are the basis of the bulk of relationships in the digital world. Digital connection does extend the scope of social connection, but not as greatly as was originally thought. Nevertheless, social-networking sites do increase the intensity or frequency of contact between people and this will continue to drive network traffic. This growth is likely to come at the cost of other activities, with TV consumption showing a marked decline.

A more surprising feature of social-networking services and sites is how their use is affecting web-based technologies and applications. Access to web content, for example, is now being increasingly driven by social-networking behaviour, and this may displace the use of third-party tools to access and navigate the web, such as search engines. Hence, the threat for, say, Google does not come from other search engines (for example Microsoft's Bing), but from social-networking companies such as Facebook. Similarly, social-networking sites are also acting as the messaging portal for their members. Users of these sites have most if not all of their friends on the same sites and hence do not need to use messaging applications outside those environments. Microsoft's Outlook email tool, for example, is under threat not from other email tools, but from social-networking sites. This will also have implications for technologies like SMS and other messaging formats currently supplied by network operators. As social-networking services begin to appear on mobiles, the users will be less likely to use voice calls, text messages, etc. to keep in touch with their social networks, and will instead simply click on the social-networking provider to gain access to messaging formats. The kind of networking in which the parties might not know each other, the second of the types listed above (such as is the case for many users of the Twitter social network and for most users of traditional blogging technologies running through PCs), will evolve in similar ways and will have similar technological and business implications. Instead of using search terms to find content, users of Twitter rely on RSS feeds, for example. Messaging will be undertaken indirectly via web logs rather than via tools expressly designed for person-to-person messaging.

New social connections made through the use of web-only experiences, such as on-line gaming, have shown remarkable growth in recent years. This has a number of implications beyond those already mentioned. Data volumes increase very rapidly with on-line gaming, and the QoS levels that are required are also very high. Latency can be a particular problem with multiplayer real-time-distributed games for example. Network providers who can achieve the requisite levels may find themselves at a substantial competitive advantage, but they may also find that the associated costs might not be so easily passed on to the consumer. Consumers might not think games warrant high traffic costs. New forms of interaction with games, particularly those that emphasise telepresence-type experiences (such as Microsoft's Natal), will also be inhibited unless the speed and rate of data access are improved and latency reduced substantially.

Despite the unexpectedly high take-up of all types of social-networking technologies, their usage has not yet appeared to reduce the volumes of older communication modes. Voice calls are still important and texting is used as widely as ever. The newer communications systems have simply extended the scale and scope of communication. For the user they have provided additional ways of interacting with others that fit alongside the older types, with people selecting the method most appropriate to the situation. For businesses supporting this traffic, this has justified the huge investments and costs that they have incurred in improving the networks sufficiently to allow these new practices to emerge. (For an excellent discussion of this see [1].)

A different kind of social communications opened up by mobile communications is the use of location. Applications such as 'find my friends' have been possible for some time, although they have not yet taken off because of the difficulty in consistently accessing location information from mobile handsets and the reluctance of mobile operators to provide this information to others. On the iPhone, which overcomes many of these problems by using a single-format device and determining its location itself, there have been a range of applications such as 'find all the people interested in a particular hobby within five miles'. As devices become increasingly able to locate themselves, or perhaps as operators make location information more widely available, the range and number of these sorts of applications will certainly increase. Chapter 15 looks at location and considers some of the applications that might emerge. With much of social networking based on telling other people where you have been and what you have done, some application that can pre-populate the 'where I have been' field might be of value.

Technology can simplify the communications process in some cases. For example, if someone watches a programme they enjoy then they can use Twitter, Facebook, texting or another method to tell their friends. Alternatively, technology could simply enable them to press a button on the TV remote control (or on the IPTV system) such that they could rate the programme and, if the rating was high enough, information would automatically be sent to a pre-defined set of friends. As this information was sent to individuals it could be consolidated, so that someone would be informed only if at least a certain number of their friends had rated the programme above a certain level. In this case, the programme could be automatically downloaded to the PVR or mobile device according to their preferences. Indeed, this process has already started with websites offering ways to inform others if a reader thinks they might be interested.

One of the more extreme concepts in social communications is the 'life camera'. This device, which has many names, is a video camera worn by an individual that continually captures all that they see, say and hear. All this information can be archived to provide a complete record of a person's life and could also be shared with friends and others. While storage is evolving to make such an application possible, the time others have to watch it is not. Selecting the interesting parts as well as ensuring appropriate privacy is one of the largest obstacles in making the life camera an attractive service. Researchers have tried some initial mechanisms for reducing the amount of data, for example triggering the camera only when a change occurs, such as movement or a change in lighting conditions. Further research and trials may throw up filtering approaches that better meet requirements.

With such a device, users might post a video clip of something interesting that happened during their day or might automatically produce a short montage of pictures from the day for others to browse. The video might also be useful in crime prevention and news gathering – effectively there would be millions of CCTV cameras available. Some users might be happy to let selected friends 'tune in' to what they were seeing at any point, which could be a useful tool for parents and carers.

The technology for such a device is already available – many mobile phones have video-recording capability and memory sizes are large enough to store a whole day's worth of video. There is sufficient processing power to implement most filtering approaches that could be imagined; however, there remains much work to do in developing algorithms that reduce the time needed to watch the video while still retaining the interesting parts.

14.3 Prognosis

Social communications is certain to grow and evolve over the coming years, although predicting exactly how is very difficult. The trend towards increasing use of mobile devices to update and access social-networking sites will certainly continue as devices and services (including messaging formats) are tailored towards making this easy, increasing the data loading on cellular networks. Gaming is likely to increase too, with better graphics engines on smart-phones and greater data rates on cellular and WiFi networks enabling richer game experiences. Location is also likely to become an increasing part of social networking, bringing those who already know each other together in new ways and affording new types of social connection. Gaming, too, may well benefit from new location-based formats.

From a wireless viewpoint, devices and networks are already able to accommodate wireless social networking, with many of the systems, such as Twitter, placing very little additional loading on the network. The major exception to this is on-line gaming. Social communications might change our relationship with our mobile devices as they become ever more important, but we will probably not see a material change in the technology or networks used for wireless communications. Only if there is a strong requirement for high-performance gaming (and a willingness to pay) or for wireless uploading of large amounts of video content will technologies and tariffs need to be modified.

Reference

[1] R. Harper, *Texture*, MIT Press, 2010.

15 Location

WITH RICHARD HARPER†

15.1 Defining location

Location is broadly about knowing where someone or something is. It may be about knowing where you are in order to navigate to somewhere else or find something. Or it may be about knowing where other people are, for all sorts of reasons. At different times differing types of location information are useful. For example, when using a car satnav the precise position of the car in terms of its geographic coordinates is important. In another situation it may be sufficient to know whether a person is at home or not. Sometimes location is relative, such as wanting to know how far away the nearest restaurant is. Sometimes it is helpful to know also in which direction a person is orientated, for example in order to provide them with information on the building which they are viewing. Sometimes precision is important; in other situations location within a kilometre or two may be adequate. Hence, understanding location is not just as simple as assuming that once a GPS fix has been achieved the problem is solved.

This section looks first at the different methods which a device can use to locate itself or be located by networks. Then we discuss some of the location-based applications that might be of interest. Finally, we look at why many of these applications are not widespread and note the implications of this for the wireless industry and other applications.

† Richard Harper <mailto:r.harper@microsoft.com> is a Principal Researcher at Microsoft Research in Cambridge and co-manages the Socio-Digital Systems <http://research.microsoft.com/en-us/groups/sds/default.aspx> group. He is the author of ten books, including *Texture: Human Expression in the Age of Communication Overload.*

15.2 Methods of location

While GPS is the most common method of location, there are many different methods of locating devices. GPS works by examining the relative time of arrival of radio signals from a number of satellites. Because the positions of the satellites are known and their timing is highly accurate, trigonometry can be used to determine position. The GPS reception process starts by acquiring a satellite signal. This can be difficult because the signal levels are very low – satellites do not have much power and are far away. The signal received by devices is below the noise floor but, because the transmitted signal is repetitive, correlators in the device can effectively increase the strength of the received signal. However, the right timing for the correlation process has to be found, and this can be done only by working through the range of possible timing positions until one works. Once the signal from one satellite has been detected, information broadcast by the satellite on its position and the positions of other satellites can be downloaded. Finally, signals from the other satellite can be acquired and a position determined. The complexity of doing this explains why GPS acquisition can often take a number of minutes, especially when the signal is weak.

GPS is generally an effective location system. It is sufficiently accurate for most applications (generally to within 10 m), it is universally available and chipsets are cheap. Its downsides are that GPS receivers require relatively large amounts of battery power (compared with, say, a mobile phone in idle mode) and that it works poorly, or not at all, indoors. It may be inaccurate in dense urban areas where street canyons mean that the visibility of the sky is obscured. GPS is likely to be augmented by other satellite location systems such as Galileo in the next decade, and these will increase the accuracy and the number of available satellites, further improving the system.

Another method of location is to make use of network identity information. Systems listen for the identity codes of nearby WiFi base stations and report these back to a network-based database, which has been pre-populated with the locations of WiFi base stations known from earlier surveys. Because WiFi hotspots cover only a few hundreds of metres this is reasonably accurate, although there may be no hotspots within range or the names of local hotspots may have changed. WiFi location has the advantage over GPS that acquisition of local WiFi identities can be performed in a few tens of seconds, but the accuracy is lower. Nevertheless, especially in dense urban areas, accuracy may be adequate for applications such as determining which friends are nearby. A similar approach is made possible by listening to the identity of cellular base stations or broadcast towers, although, since these tend to have larger cells, the accuracy falls as a result, perhaps to 1 km or worse.

In some parts of the world there is another location system known as Loran. This uses ground-based transmitters working at relatively low frequencies to send pulses at precisely timed intervals. As with GPS, by listening to the relative arrival times of these pulses, receivers can deduce their position. Loran was designed for shipping use and hence is most accurate over the sea. However, it can be used over land and does have the advantage that its signal can penetrate buildings relatively well. It is less accurate than GPS, with an accuracy of perhaps 100 m or so (it is not clear just how accurate it might be in urban areas), and requires dedicated receivers and antennas in the device, which can tend to be rather large for handheld devices. Its future is also unclear since its long-term funding is in doubt.

A different approach is for the network to determine where a device is. In this case, rather than the device listening for signals from multiple transmitters (e.g. satellites), multiple receivers in the network (e.g. base stations) listen for the signal transmitted from devices. There has been much work on this in order to determine the location of cellular phones, especially when making an emergency call. Approaches broadly fall into those employing angle of arrival (AoA) and time difference of arrival (TDoA). With AoA, antennas at the base station locate the direction from which the signal is arriving. If multiple base stations determine the direction from which they receive the signal then simply noting where the lines of bearing cross gives the location of the device. This approach is little used because expensive direction-finding antennas are required and because, in an urban environment, the signal may arrive via reflection from a direction different from that corresponding to the location of the device.

TDoA is simpler to implement, with each base station listening for particular features in the signal and then comparing the times at which they arrive at nearby base stations. No special antennas are required, although base stations do need to be accurately time synchronised relative to each other. Some networks do now use variants of TDoA to locate mobiles. Its accuracy depends on the size of cells and the number in range of the mobile – in some cases there will be only one and no location is possible (other than knowing that the mobile is in the cell). In best cases accuracy to within 50 m is possible.

15.3 Cases where location is difficult

Two of the technologies discussed earlier require location to function well but have problems – these are femtocells and cognitive radio.

Femtocells are located in homes. Operators need to be fairly sure about where they are located. In the worst case, if a femtocell were taken to another country, it might start transmitting on the spectrum owned by a different operator, causing interference. Also, accurate location information allows the operator to undertake network planning and to build appropriate handover lists. As a result, it is valuable for the femtocell to report its location to the operator when it is connected to the network and if it is ever moved. GPS would be ideal for femtocells, providing a high level of accuracy. However, GPS works poorly indoors and might not be available. Some early femtocell manufacturers provided GPS antennas on long leads allowing the antenna to be placed near a window while the femtocell was deep inside the house, but users disliked the need for leads and this approach no longer appears to be used. Another approach is to take much longer to acquire the GPS signal using longer-length correlators. This can be assisted by information from the cellular network as to the likely timing of GPS signals (so-called assisted GPS – A-GPS). This may work for femtocells indoors but near the edge of the house. If these fail, then providing information on the base-station identifier of nearby cellular base stations may be sufficient to show that the femtocell is located approximately in the region where it has been registered. Some operators appear to have decided that it is sufficient simply for the user to self-register their location.

As mentioned in Chapter 7, cognitive devices may work best if they can locate themselves and then request information on available channels from a database. While it is not clear what cognitive devices will be used for, at present most cellular and unlicensed use is indoors. As has already been mentioned, GPS functions poorly indoors, making it difficult for the cognitive device to determine where it is. Since the cognitive device may have been carried about, one solution is 'GPS-lock', whereby the position of the last fix is remembered. So, if the device were carried in from the street to a house where there was no GPS coverage, the device would record its position as that of the street outside the house. This would be of adequate accuracy for cognitive operation. However, there is no guarantee that the device has not been moved from one house to another, perhaps without being switched on, and hence may be in a totally different location. Some mix of using the identities of nearby WiFi and cellular base stations may help, but such information might not always be available. Lack of guaranteed location information may be a serious problem for cognitive devices, and solutions to this are actively being investigated at the moment.

15.4 Beyond location

Existing location services are broadly based on determining where the user is, where other users are and, if required, providing means of getting from one to the other. They essentially provide a better map. Proposed location services move beyond this. They are of two types. We deal with each separately.

Micro-location

The first type of LBS, which is already commonplace, uses local area networks, created on an ad-hoc basis, to message to the passive mobile user. Bluetooth is used by advertisers to send unsolicited content to nearby users. This is called proximity marketing or Bluetooth marketing. It is analogous to users deploying Bluetooth to send content to other, nearby users, which is commonly referred to as 'bluejacking'. Whereas bluejacking is often thought of as mischievous, almost invasive, Bluetooth marketing is typically viewed as more acceptable, though clearly this depends on the kind of content delivered. Currently Bluetooth marketing is used by advertisers in high-density settings, such as underground stations in London, gateways to shopping centres and restaurants.

There are important implications from the emergence of proximity marketing. To begin with, it is a novel way of using Bluetooth. The protocol was devised around the idea of pairing, with the expectation that the user would maintain control over the transactions, but this form of content production pushes content to the user without pairing being required. Also, this technique does not require user permission, at least not in advance. As it happens, some devices require the user to accept the delivery of a file whereas others do not. It is also the case that the out-of-the-box settings for Bluetooth vary with device and OS supplier too, and some of these require the user to preset the device to allow receipt of pushed content.

A further feature of Bluetooth marketing is that it is very local. Bluetooth marketing is often possible only within a few feet of a transmitter. But it is precisely this that appeals. Advertisers can guarantee that the user is engaged with certain concerns, such as an advert on a wall; while owners of restaurants can be sure that the user is in their restaurant or, at worst, walking by. Recipients of Bluetooth marketing also seem willing to accept such content since their location justifies it – or at least makes it seem relevant. In this regard, Bluetooth marketing highlights the fact that some examples of LBS are micro-local. Such services add value in the social contexts in which they are experienced – restaurant price offers are delivered to those in the restaurant, for example.

Micro-location values are different from those created by broader-grain location information. New values might be uncovered in the future. To date, the wireless industry has not properly investigated how the micro-local might be developed; nearly all the current deployments have been developed by third parties such as marketing companies, rather than by the industry itself. This indifference seems systemic in the wireless industry – for example, the planned future releases of Bluetooth protocols do not reflect these usage patterns.

This transaction method is free to both the recipient and the Bluetooth marketer. The wireless networks gain no revenue from these transactions. Indeed, it is difficult to see how any revenue could be directly gained. Nevertheless, the fact that the provider of micro-local content perceives value, just as the recipient does, suggests that there are benefits to the overall ecology that the wireless networks and infrastructures provide.

Augmented location

Micro-local LBS draws attention to how the virtual and the real are being brought together in ways that were not always predicted by the industry. Another form of LBS brings the user, via the mobile, into the world of virtual reality in even more powerful ways, but here the wireless industry has shown more interest.

The departure point for a range of new services is a form of augmented view. Here the user holds their mobile device like a camera, pointing towards a view or object that interests them. Using GPS, the mobile device works out where it is, and, using an embedded compass (a necessary feature for this service), it works out the direction in which it is being pointed. A camera in the front of the phone relays an image of what the user is looking at onto the phone screen, so the phone screen appears to the user to be a window through which they are looking onto the view. Additional information can then be superimposed on the image by the phone, such as the names and heights of mountains, the review ratings of restaurants in view and so on. This information is retrieved from databases that know the location of each point of interest and can deduce where that will fall within the view of the user. Figures 15.1 and 15.2 below show screen shots of a real estate application showing information about homes for sale superimposed on the video feed and of mountains with a peak identified.

An enormous number of applications for this augmented view will be possible. Many are currently being introduced to the marketplace, with Google maps on mobile phones (in particular on Android devices and iPhones) being linked to location services. When a mobile user selects a geographic point, information about nearby restaurants is imposed on the map display.

Many others could be imagined, from showing a recommended route superimposed on the real world to tourist information when visitors are walking around a city. Augmented view does not have to depend on location – for example an augmented view of products in a shop could be provided if there were some barcode or RFID on each item that the phone could recognise and use to look up an appropriate database such as review comments about books on a shelf. Augmented view merely requires devices to have GPS and a compass, and to be able to download the appropriate application. The views available then

Figure 15.1 A screen image of an augmented view providing real-estate information (source: http://www.layar.com/).

depend on the information provided by entities such as Google Maps and many other bodies. Given the relatively rapid speed with which users replace mobile devices, it seems likely that augmented view will be adopted over the coming few years. Network requirements are relatively limited – the phone transmits only its location and direction of pointing to the network, and the appropriate website returns a small amount of information for the phone to process.

Once the phone is able to add to, or modify 'reality', it is a short step to extending augmented reality so that the virtual has greater salience in real lives. The phone could make the world appear a better place by modifying the sky. Though this will not result in the virtual replacing the real, it could alter the value that modern-day buildings, cars and other built infrastructures (rather than coded ones) provide the user. It could even be that augmented reality could

Figure 15.2 A screen image of an augmented view identifying mountains (source: http://www.layar.com/).

deepen relations between people. Applications such as presenting the user with the name and last time of meeting of anyone walking towards them could be provided on the basis of information sent from other users' mobiles, for example. Whether any of this is actually valuable or acceptable is not clear, but location makes it technically possible and existing wireless networks provide sufficient capacity and coverage.

15.5 The 'Whereabouts Clock'

It is doubtless the case that other new forms of location-based services will appear. As the Apple Apps store has shown, many different applications based on location can be envisaged, from 'find the nearest Starbucks' to 'show me which of my friends are close by'. Many more uses of location are likely to emerge over time. Yet any investigation into the history of the wireless industry would make it clear that the industry itself has not always benefited from these opportunities, despite the fact that imagining new services does not seem, on the face of it, difficult to do.

Consider, for example, the 'Whereabouts Clock'. In the Harry Potter novels, in the home of the Weasleys there is a device that looks rather like a grandfather clock but with one hand for each member of the family. Around the face of the clock are locations such as 'work', 'home', 'pub' and 'travelling' (and 'in dire peril'!). The hands move around the clock as family members travel, showing at a glance where each member of the family is located. Location is represented not as a person's actual coordinates but by which of a number of possible location categories they fall into. The clock is always up to date – just like with an ordinary clock, there is no need to press a 'refresh' button to update it. In the Harry Potter novels this clock works by magic, but it would be simple to implement it using available technology (except for the 'dire peril' setting). This section looks at ways in which such a clock could be implemented and why it is not yet available (although close substitutes are becoming possible).

The outline of such a clock would be a device in the home (perhaps the kitchen) which might be based on a tablet PC, with an always-on screen, running the location application. It would have wireless capabilities to connect to various networks. Family members would be located using the wireless devices that they were carrying.

There are two broad approaches that could be adopted to determining the location of family members – device-based and network-based. These are discussed below.

In a device-based approach, the personal device of each family member would work out its location and send information to the network or directly to the home monitor. The device might continuously check its location using GPS or it might periodically check whether it had been moved (using its internal accelerometer). It might also monitor WiFi identities, since this might be a good guide as to when it was in particular named locations such as the home. The home monitor then interprets this information and displays it accordingly. This approach is relatively simple to implement, requiring only an appropriate application for the device, but would require each user's mobile to have GPS and WiFi capabilities and to leave these running for much of the time. The mobiles would also be frequently sending location messages, which may drain their batteries and incur network charges. The Google Latitude application works somewhat in this manner, displaying the location of a preset list of people whenever the application is invoked on a computer screen. This could be argued to be already similar to the Whereabouts Clock. However, at the time of writing it ran on a relatively small subset of handsets and did not continuously update or categorise location into a short list of possibilities.

In the network-based approach, the network determines the device location from an understanding of the cell, or sets of cells (known as location areas), that the device is in. This information is already generated automatically as the device moves around the network, primarily so that the device can be located if there is an incoming call, but can also be used to provide approximate information. Coupled with WiFi identification of specific locations, this may provide sufficient accuracy. The advantage of this is that it works with all handsets and does not cause any additional battery drain. The operator can deduce that a handset has moved when its location area changes and send this information to the home monitor, perhaps directly using a cellular link. The downside of this approach is that it will require minor modifications in the network location registers in the cellular networks. It may also be problematic if various members of the household are using different cellular network operators since they might not all implement the solution, or might not do so in a compatible manner. Although some operators have made an effort to provide family location (e.g. the Disney MVNO in the USA had as a core proposition the ability of a parent to locate their children), in general this is not available.

With all location applications there are some ethical issues around tracking of individuals, but it might be thought that, in a family where consent is provided by all, these issues would not be major problems.

So both device-based and network-based solutions could readily be implemented today, albeit with a few downsides for both. The application is clearly very valuable. Yet, no solutions are widely available. The next section looks at why this is the case, and in doing so draws out some valuable lessons for wireless applications and services.

15.6 What location tells us about the wireless industry

For many years, the cellular industry saw location-based services as a significant source of new revenue. The operators felt that they had unique information in the form of location and could charge subscribers a premium each time they used this location information. The operators offered a number of early services such as 'find the nearest restaurant'. Take-up of these services was very low, perhaps because subscribers felt that the charges were unreasonable. Nevertheless, operators persisted in believing that they had a valuable and unique source of information.

A number of application providers, such as Google, agencies that work for such providers (for example advertising and marketing companies) and device manufacturers such as Apple, finally decided that it was simpler and more likely to be profitable for them to circumvent the operators and seek alternative sources of location information. This coincided with the increasing availability of Bluetooth-enabled handsets and GPS in handsets, as well as downloadable applications and WiFi hotspots. As a result, cellular operators have broadly lost out. Location-based revenue-generating concepts have mostly accrued to search companies such as Google, while micro-local experiences that enhance the perceived value of the mobile infrastructure have been developed by companies and agencies that are mostly outside the wireless industry. Perhaps the wireless industry has had an inflated view of the worth of LBS and their role in the process of delivering such services; perhaps the industry has also neglected to appreciate the technologically simpler mechanisms for producing value that the infrastructure they have developed enables. This is certainly the case with many micro-local services. Beyond all this, it may be that an excessive confidence in monetisable value and technologically elaborate systems has led the wireless industry to over-play its hand.

If it is the case that attempts to introduce location-based services by the operators mostly failed, it also the case that the operators (and indeed the terminal manufacturers to some extent) have failed on other areas of service innovation too. Here are some examples.

- Attempts to introduce email services failed until Blackberry arrived, effectively bypassing the operators by installing devices in the corporation and software in the handset and just using the data facilities of the operator's network.
- Attempts to introduce Internet access using 'walled gardens' such as Vodafone Live gained little traction; it was the introduction of the iPhone and associated applications (not just a browser) which translated any web page into a form ready for use that revolutionised the mobile internet – again using only the data facilities of the operator.
- Although it is early days, it appears that watching mobile TV is more likely to occur via a download from the iPlayer of a podcast suitable for a mobile device than through anything offered by the operator.

Indeed, it is hard to think of a single example of a service that the mobile operators have introduced that has become successful, despite many years of trying, not just with the examples above but including also picture messaging, home-zone tariffing, push-to-talk and other user-group services, mobile payment, music services and so much more.

What the mobile operators have done well at is the provision of 'bit pipes' – basic carriers that transfer voice and data from one place to another. Voice and SMS have been hugely successful and more recently 3G data is starting to take off now that prices have fallen and operators are concentrating on bit-pipe provision.

So there is something of a theme emerging here. Operators are very good at providing voice and data transfer but very poor at delivering services to run on top of these. This is despite their fear that they will be marginalised if they become just a bit-pipe provider and their understanding that the only way to continue to grow and be profitable is to 'move up the value chain' into service provision.

The reasons why operators have failed so conspicuously to deliver services are manifold. Firstly, services that exploit the wireless infrastructure but are not related to person-to-person communication are not their core expertise. For example, though they have a great deal of expertise on location and its relevance to the delivery of high-quality voice and SMS communications, they have less understanding of how to convert that knowledge into insights about location-based services. When the operators move beyond 'being in touch' they seem to lose their touch. Companies and other organisations that have developed a business on the basis of services other than communications have done much better – such as Google, which has used understanding of search on the web as an analogue to search in the real world, and hence successfully integrated mapping data and location into much of what Google does.

Nevertheless, just as Google has managed to stretch its competence into new areas, so it is also possible that the operators could do so in the future. This may require a bold shift in their hiring policies and organisational structure to give greater prominence to non-communications engineering skills and know-how.

Secondly, and perhaps related to the first point, the operators (and wireless industry more generally) may have to accept that technological enhancements or technological efficiency might not guarantee better user experiences or a flourishing ecology of services. Just as Apple had to devise an apparently poor OS (one that can handle only a single application at a time) to ensure that the user experience it wanted to deliver on the iPhone was good enough, so it might be that some of the services that the wireless industry could develop will need to be developed around technologically prosaic or even archaic systems. The evident (albeit modest) success of Bluetooth marketing is proof that more powerful LBS networks are not the guarantor of experiences that users will take up. By the same token, the micro-local and essentially ad-hoc networks that Bluetooth enables may seem unappealing to an industry that assumes that richer and more accurate location information and networks will produce better services.

The lesson from this might be that it ought to be the services that determine network provision and device function, even if that ends up requiring networks and devices that are less technologically advanced than engineers can deliver.

Thirdly, the operators may have been distracted by expectations that have not fulfilled themselves. The idea that mobile and fixed devices would somehow converge, for example, is now no longer believed, even by companies that have much to gain if it were to happen. Microsoft, for example, worked on the principle that the mobile and the PC would eventually merge and so designed its OS for the mobile as a replica of its OS for the PC. Only now has it abandoned this idea and switched to developing a mobile-specific OS.

Fourthly, the mobile industry has not shown itself capable of developing good relations with other companies in ways that ensure business success for all. Even amongst themselves they have a history of difficulty. For example, the ideal Whereabouts Clock would probably be based on a mix of information provided by operators and from devices. It would be flexible, depending on the type and capabilities of the devices being carried by family members. It would work even if family members used a range of different operators. However, experience suggests that the industry working together in this manner and being prepared to share location information is very unlikely. It is not clear how this might be altered, except through a change in culture.

Fifthly, it may be that they are trying to extract more revenue from the service than is viable, or than consumers are prepared to pay. In so doing the industry may have failed to recognise that overall business might be enhanced by providing a richer ecology of experiences to the user. The World Wide Web took off not because any particular service or website offered huge value, but because large numbers of websites offered value in the aggregate. Besides, revenue can be generated in ways other than traffic – search-engine providers do not make their money simply from the number of visitors they have; indeed, the user of a search engine typically does not pay anything for using it. It is what the use of a search engine leads the user to that generates the money for search-engine providers.

Sixthly, mobile operators do not have the right image with consumers. While they have a very strong brand, it is associated with being a bit pipe – with the provision of voice and data to a mobile phone – rather than with innovation, with being cool or even with being a trusted entity. That is why individuals are much more willing to try a new service from Apple or Google than they are with one from Vodafone or Orange. Equally, they probably would not want Apple to deliver the voice service that they rely on as a core part of their life. Brand can be critical in these areas.

This is a conundrum for operators. They want to avoid at all costs becoming a bit pipe because they perceive that this would result in lower profitability and growth, and yet they are extremely good at bit-pipe provision with a brand and organisation well matched to delivering this. Operators would prefer to deliver services, but their track record is awful, they are routinely out-manoeuvred by organisations outside the mobile space and they do not have the right brand or skills to achieve service delivery. If operators really wished to deliver services then perhaps they should split themselves into bit-pipe organisations and service organisations, enabling separate branding, skills and focus, but it is hard to see this happening.

The implication is that the development of services will probably be done by others, such as Apple, Google, Microsoft or even entities such as Amazon. By working with these companies and making various network parameters available, operators could stimulate demand, leading to great bit traffic and hence increased revenues. However, it is likely to be many years before operators realise and embrace this, and until that time application providers will have to work around operators rather than with them, to the detriment of services like the Whereabouts Clock.

16

Healthcare

WITH RICHARD HARPER[†]

16.1 Introduction

Healthcare is an area of rapidly growing importance. Many studies have shown clearly how the population is ageing, needs for healthcare are growing and yet the number of people in work (and hence paying for the healthcare system) is falling. Under current extrapolations, healthcare will increasingly become unaffordable and impossible to implement, with an unfeasibly large percentage of the population engaged in caring for others. Either the quality of healthcare provided will fall or new means must be found to care for those who currently rely on the support of others.

Wireless communications provide one possible part of the solution. Through a system of monitors, alerts and the provision of information, it might be possible for electronic systems to allow people to monitor their health more effectively or to generate better information that remote medical professionals can monitor and analyse. The benefit of both scenarios is that the users of wireless medical infrastructures would be able to stay in their homes and look after themselves for longer. A downside is that the increasing medical monitoring will uncover mild forms of sickness and ailments that have hitherto remained invisible, or at least have been dealt with by the natural defence mechanisms of the body without medical intervention. Another, related to this, is making individuals excessively health conscious, hypochondriacs in other words.

Developing wireless solutions to healthcare needs might be important both to society as a whole and also to the wireless sector, where important new markets

[†] Richard Harper <mailto:r.harper@microsoft.com> is a Principal Researcher at Microsoft Research in Cambridge and co-manages the Socio-Digital Systems <http://research.microsoft.com/en-us/groups/sds/default.aspx> group. He is the author of ten books, including *Texture: Human Expression in the Age of Communication Overload.*

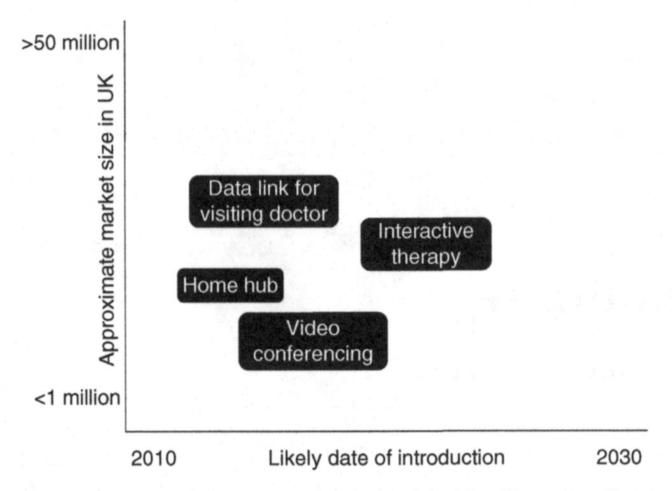

Figure 16.1 Possible communications-related healthcare applications in the home (adapted from [1]).

and new uses of technologies might be enabled. This chapter considers the possible roles of wireless in healthcare, drawing on the technology discussions in the previous part of this book. It divides healthcare into the areas of the home, hospital, ambulance and individual.

16.2 Healthcare at home

A summary of possible developments of healthcare in the home is provided in Figure 16.1. Broadly, the functions of devices in the home are as follows.

- Monitor the individual to make sure that there is no cause for concern.
- Provide information and remote assistance when problems are detected.
- Enable communications for visiting healthcare professionals.

Monitoring

Monitoring can take many forms, from fall alarms that detect a sudden change in patient orientation, to sensors in pill-dispenser boxes that detect whether the daily dose of pills has been taken. Other sensors could be included in scales that chart daily changes in body weight and composition, or movement detectors to assess whether the patient is moving freely around the house.

Such monitoring essentially forms a sensor network of the sort that we have discussed in Chapter 13. The sensors may be wired or wireless – for the most part wireless sensors would provide more flexibility and avoid the need for complex wiring systems to be inserted around the house.

One of the key problems noted for sensor networks was the powering requirements, but in the home these may be eased. Some sensors might be able to use mains powering and in others (e.g. scales) sufficiently large batteries could be employed to provide many years of power. Some sensors could have their power supply replaced regularly, e.g. sensors in pill-dispenser boxes might have a new battery inserted each time a new supply of pills was delivered. Because most of the sensors will be readily accessible, replacing batteries should they fail unexpectedly might not be problematic (as long as a failed battery was identified as the fault).

Because most homes are relatively small environments, a communications system based around a star architecture might be appropriate. This would consist of a home hub providing wireless communications throughout the home. The most likely candidate for this would be WiFi, which already provides such functionality, although using WiFi in the sensors might be a heavy drain on batteries. Nevertheless, as noted above, this need not be problematic in many situations. WiFi also has the advantage of inexpensive chipsets that can readily be built into many devices. Alternatives could be cellular-based (using femtocells), Bluetooth, ZigBee or proprietary protocols. Mesh architectures could also be considered. Any of these could be made to work, but WiFi has the big advantage of the home hub already being available in many cases. Another approach might be a number of powered distribution nodes around the house that communicated with the home hub via WiFi or home powerline communications and communicated with sensors using a lower-power and lower-range standard such as Bluetooth.

The sensor information could then be processed locally by a home PC or similar device with alerts sent to the health authority over the broadband connection as required. Alternatively, all the data could simply be passed directly to the health authority for processing and storage in a central server. Both approaches can readily be implemented and perhaps both options will be available.

Providing information

This is simply a means of allowing patients to describe their problems or symptoms and receive appropriate advice. It can be provided using an Internet-style information portal and merely requires the patient to have a PC or similar and preferably a broadband connection, although narrowband connections might be adequate in some cases.

Enabling communications for visiting healthcare workers

This is also relatively simple. A visiting doctor or nurse could connect to the home hub, most probably using WiFi, and then use standard networking arrangements to link into central databases or other information sources.

Methods of introducing a home healthcare system

A key question is how such a home healthcare system could come about. Broadly, it could happen in an evolutionary fashion as home owners gradually added various items one at a time as they found them valuable or, alternatively, it could be offered as a complete 'home refit' with an entire system being installed.

The complete refit offers the possibility of a single system, optimised for the task in hand, without the complexity of having to adapt to the various elements already in the home. It could be installed by professionals and might be simpler to use than multiple different systems because it could have a single combined user interface for all sensors. However, this would require central planning and probably central funding. While this is not out of the question, there are many barriers to such deployments. These include issues such as the department benefiting the most (often the hospital) not being the source of funding (which is often the local authority). While it should not be ruled out, it seems unlikely.

Evolutionary deployment is much easier to envisage. A simple example of a home healthcare element that could readily be implemented now is a set of scales that measure weight and body-mass composition (these are widely available for \$50–\$100) with inbuilt WiFi, which recorded the daily measurements and then supplied them to the home computer via the home hub. Such a set of scales could probably be provided for less than \$100. If the home already had a WiFi system and home PC then, with the download of a simple software program, the scales could provide information both on the weight of the patient and also on the fact that they had weighed themselves at the usual time of day (and hence were active). Indeed, this functionality is already available from the Nintendo Wii-Fit, which uses its balance board as a set of scales and then displays information on weight change graphically. It has inbuilt WiFi to connect to the Internet and can download applications that would allow tailored functionality.

There might be a gradual increase of such sensors over time, with end users buying them directly, making use of subsidies where available or building on functionality already provided in devices such as the Wii. Professional assistance might be occasionally required in order to install new elements or update software to handle the diversity of devices in the home, but this is little different from what already applies to homes of today, which have multiple audio-visual devices and are increasingly acquiring sensors such as security systems.

16.3 Healthcare in the hospital

A summary of possible hospital healthcare applications is shown in Figure 16.2. Broadly these fall into the following categories.

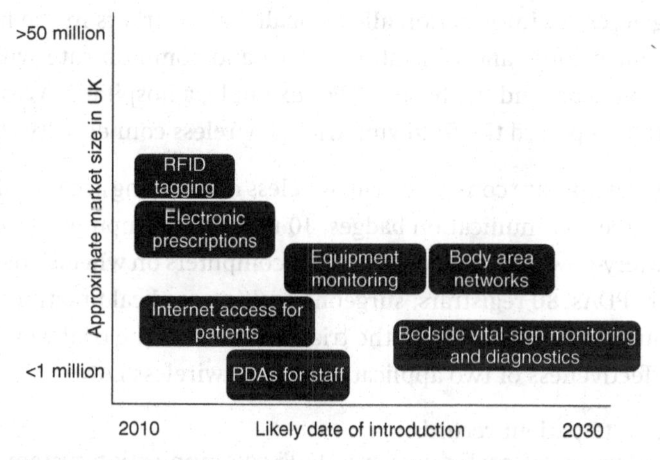

Figure 16.2 Possible communications-related healthcare applications in the hospital (adapted from [1]).

- Monitoring the location and movement of people, equipment and drugs.
- Removing wires.
- Facilitating access to information.

The monitoring function is valuable because items can go missing in hospitals. Valuable equipment can be moved by one person, resulting in their colleagues being unable to locate it later. Drugs can be misplaced, or given to the wrong person. Even patients can go missing if they choose to move around without informing the staff. Tracking these items can save time and costs, and in some cases could result in better care.

Tracking is generally performed by placing a passive sensor or tag onto devices or in the wrist bands worn by patients. Whenever these devices pass through a controlled point, such as a doorway where an active reader is located, the reader notes their presence and records it. Such tagging is already widely used in department stores at the store exit to detect stolen goods and in business premises to open doors with an employee pass. The technology is often known as RFIDs and is well established. Installing it in a hospital would involve some cost but little else.

The removal of wires allows patients with body sensors to move about freely and means that equipment could be wheeled from bed to bed without needing to be plugged in. This is all relatively straightforward, making use of either short-range standards such as Bluetooth (for example, to connect body sensors to nearby monitors) or longer-range standards such as WiFi (to connect monitoring equipment to central databases).

Facilitating access to information allows healthcare workers in the hospital to access patient records and central databases and communicate with each other while moving around the hospital. For example, a hospital in Westmead, Sydney, Australia reported the following trial of wireless connectivity [2].

> The infrastructure consisted of 40 wireless networking access points, 40 hands-free communication badges, 10 notebook computers, operating on battery-powered trolleys known as 'computers on wheels', or COWs, and six PDAs. 80 registrars, surgeons, visiting medical practitioners and nurses were involved in the trial. The aims of the trial were to test the effectiveness of two applications over a wireless LAN:
>
> - access to patient records;
> - rapid voice connectivity over a Wi-Fi communication system.

The trial proved effective, increasing productivity and decreasing the time taken to provide care. From a wireless-connectivity viewpoint the connectivity is simple and can be readily provided. It is perhaps surprising not that such concepts are being trialled but that they are not already widely available.

One concern over implementing these applications using wireless is that most envisage using unlicensed spectrum, often at 2.4 GHz. There are no guarantees of freedom from interference in this spectrum, although it might be possible in a hospital environment to control the devices within the hospital to some extent. If the hospital were to become reliant upon unlicensed spectrum to the extent that it could not function properly without such wireless connectivity then the consequences of interference might be severe. Indeed, the possibility of congestion or interference is one that potentially affects much more than hospitals, including WiFi hotspots, home hubs and much more. It is still very much an open question whether interference will occur and, if it does, whether there will be acceptable solutions.

16.4 Healthcare in the ambulance

A summary of possible healthcare applications in and around the ambulance is shown in Figure 16.3. These applications are somewhat limited relative to those in other sections and broadly involve either

- providing a wireless hub at an incident to connect to wireless sensors and mobile healthcare devices or
- linking back to the hospital to access information and provide information on the incident to the hospital.

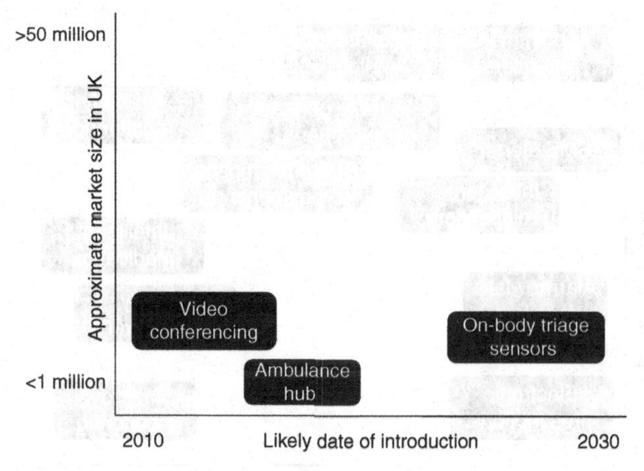

Figure 16.3 Possible communications-related healthcare applications in the vicinity of the ambulance (adapted from [1]).

Both can be achieved with existing wireless technology. The provision of a hub is simple, although it would need to use a harmonised standard such that all the sensors were able to communicate. This might mean defining the technology employed in sensors routinely worn by individuals (in which case an open standard such as Bluetooth would be appropriate). Even if this is done, Bluetooth devices require 'pairing' before they can work together, but doing this might not be possible in an accident environment. Further work might be needed to allow 'over-ride' in such cases.

The link back to the hospital is achieved using the best available communications network, which will probably be cellular or a dedicated emergency-service network.

16.5 Healthcare for the individual

Applications for the individuals are those which the individual either takes with them or accesses regardless of location. A heart-rate monitor is a good example of such an application insofar as it is linked not to a location but to an individual. A summary of possible healthcare applications for the individual is shown in Figure 16.4.

Individual applications can be broadly characterised as

- body-based sensors monitoring vital signs,
- drug-dispensing systems,
- reminders and
- access to information.

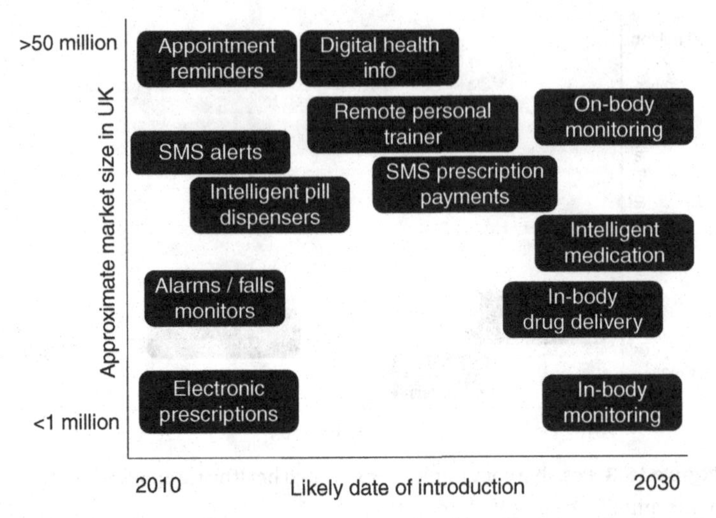

Figure 16.4 Possible communications-related healthcare applications for the individual (adapted from [1]).

Many of these are similar to applications that have been discussed earlier. Body-based sensors, such as heart-rate monitors, are already in use and as sensor technology improves their usage can be expected to grow. It would be helpful for all sensors to use the same wireless technology and to interwork so that, for example, someone wearing multiple sensors would need only one monitor (such as a wrist-worn unit, or a mobile phone) that could receive the output from all the sensors and process it accordingly. At present, many such devices have a proprietary wireless link protocol. Bluetooth might be an appropriate technology to standardise on, especially since it is already available in most mobile phones.

On-body drug-dispensing systems might be developed that could, for example, take the output from a blood-sugar-level monitor embedded somewhere on the patient and modify an insulin dose accordingly. The connection between the monitor and the dispenser might need to be wireless, not just for convenience but perhaps because the sensor is embedded within the patient. While technically this is just another short-range wireless link, it is a highly critical one. Interference could be life-threatening. For that reason, most conclude that spectrum dedicated to healthcare applications should be used for such systems. The fact that this would make interoperability with other sensors and healthcare system difficult need not be problematic since such dispensing systems are likely to be standalone.

Reminders help patients remember to take medicine or to attend out-patient appointments. They are already provided as text messages on mobile phones and further evolution could be imagined, for example, as an application on a device like the iPhone.

Access to information enables patients who are away from the home whose condition has changed to find out what they should do about it. It is effectively just Internet browsing on a mobile phone or similar.

16.6 Difficulties in introducing new solutions

On the face of it, healthcare would seem an excellent place to introduce new wireless services and technologies. There is a strong need for new solutions and many worthwhile concepts that would improve the lives of patients and save money for healthcare authorities can be envisaged. Some of these services might then be further developed into other areas of our lives.

However, there are barriers caused by the manner in which healthcare is provided in many countries of the world. Institutional arrangements are often such that there is little budget for new ideas and the benefits often fall to budget holders other than those who would fund the deployment. Healthcare is also often a conservative world that is difficult to change, both because of the entrenched self-interests and also due to the risk of harming patients, or of facing legal challenges if a deployment is not successful.

In addition to this, the benefits that continuous monitoring provides might not be as great as benefits derived from greater investment on more traditional forms of medical infrastructures. Health organisations may need to choose between more operating theatres and wireless systems, just as they might have to choose between wireless devices and more general practitioners.

There are many areas where harmonisation would be very valuable, for example in the standards that sensors use to communicate, so that individuals could have one monitoring device and healthcare professionals could link into the sensors when needed, but there is little evidence of harmonisation of this sort taking place (although it would not be unduly difficult since many appropriate standards exist already).

16.7 The social context of health

Though the benefits of wireless medical systems are self-evident, it is also the case that the role of these systems needs to be placed within a larger compass of concerns – health is not merely a biological problem but also a social one.

For example, in the short term, wireless monitoring systems might not enable the medical profession to improve the overall health of the population. Instead, new systems might simply serve to highlight sickness that might otherwise not have been made visible. Sometimes this visibility will be an advantage, highlighting, for example, the onset of a disease before it becomes too serious. But in other cases wireless monitoring will draw attention to ailments that would otherwise have been dealt with by the individual body's natural defence mechanisms. These take time to operate, and an individual who is made aware of an ailment may be unwilling to wait for their own body to respond. They may seek the instant comfort of immediate medical assistance. In this respect, wireless monitoring will increase sickness, not improve health. Over time, of course, people may come to learn what is best left for time to heal.

Second, although wireless monitoring will make people more aware of their medical condition, it will not necessarily also educate them to be knowledgeable enough to judge what the symptoms they perceive might mean. As the previous point highlights, a little knowledge might be harmful. Wireless systems could lead people to wrong interpretations, to worries that are unfounded, even hypochondria. A solution could be to offer further instruction and guidance to people on how to interpret the data that wireless monitoring generates, but this would be an additional cost. Otherwise, simply monitoring the variation of the body might lead some people to become hypochondriac, irrespective of their understanding.

Besides, it would be incorrect to assume that the role of medical institutions is solely to do with the health of the body. The role of the general practitioner (GP) in the British NHS is as much about offering psychological comfort as it is about providing medical intervention; and it is not simply providing a friendly face that is required. GPs can sanction one or two days away from work with a sick note, for example. This rest will not be for the body, but for the dispirited mind. Wireless monitoring systems will have no role in this if they are confined to body metrics. It might be that wireless systems could also monitor social behaviours – noting the fact that a person who lives alone has engaged in no communication for some days, for example. Here mobile systems could use being in touch as a metric for social health.

In the longer term, the widespread adoption of wireless monitoring might benefit not the individual user, but medical science itself, which thereafter will be able to provide better medical intervention. Currently medical knowledge is based on very-high-level aggregates of data where individual yet healthy variation in performance is elided. This is a reflection of the limited scale of medical trials – even though these are enormously costly in time and money, they are rarely able to encompass very large populations where data is fine-grained. Widespread use of data sources could enable the medical profession to build a more accurate picture of the spectrum of human performance. Thus monitors will become better able to indicate whether an individual's deviation from either side of a population 'norm' is healthy or whether it indicates something more sinister.

16.8 Prognosis

These concerns make it most likely that wireless healthcare solutions will be introduced piecemeal, often by individuals. It is much simpler for an individual suffering from a particular condition to purchase a sensor or device that will assist them and link it to their home system than it is for a central agency to supply appropriate technology. This may also reflect differences in understanding and capacity to interpret monitoring systems on the part of the individual. Self-purchase works better in a market economy and seems likely to happen over the coming years.

This, however, would lead to problems in some areas. For example, an individual who suffers from heart problems might wear a heart-rate monitor linked to their mobile phone, allowing an alarm to be raised with the emergency services if an abnormal rhythm is detected. However, unless the device and its diagnostic software have been approved by the local health authority, they may be unwilling to commit resources to taking the individual to hospital for what may be a false alarm. This suggests that it will be simpler to install basic monitoring equipment that might lead to, say a doctor's appointment, rather than equipment that leads to an emergency response (although the latter might be more valuable in terms of saving lives). Here the relationship between biological health and the social context of health provision is made clear.

Irrespective of the balance between the social and the medical, a bigger lesson for the wireless industry is that much of this comes back to the value of a home hub that can provide links to the sensors around the home. This might well be WiFi, possibly supplemented with subsidiary WiFi/Bluetooth convertors in strategic locations around the home. WiFi might also be widely deployed in the hospital. Both of these show the increasing value that might be derived from unlicensed spectrum but also the risk that this might suffer from interference or congestion both of a technical and of a social kind.

References

[1] http://www.ofcom.org.uk/research/technology/research/sectorstudies/health/.

[2] M. Miller, *Evidence-Based Health Care Connectivity – Putting the Promise back into Technology*, 2007. Retrieved September 7, 2009, from Health Informatics New Zealand: http://www.hinz.org.nz/journal/2007/10/Evidence-based-health-care-connectivity-putting-the-promise-back-into-technology/972#10.

17

Transport

17.1 Introduction

Transportation is an area where wireless can provide important services. From warning of congestion through to automatically guiding vehicles there are many benefits that wireless could bring. Wireless already plays a substantial role in some sectors – for example, air travel without wireless communications and radar is hard to imagine. But in other areas, such as driving, wireless plays a more limited role of providing radio entertainment and satellite navigation.

Transport is perceived to be one of the largest contributors to greenhouse gases and there is much interest in reducing emissions. Wireless could potentially play a role by making transport systems more efficient.

This chapter looks at each of the key modes of transport and considers the role that wireless might play and the difficulties in its introduction.

17.2 Road

Road applications can be divided into

- route guidance,
- safety and
- vehicle telematics.

Route guidance

Many now make use of satellite navigation (satnav) systems to guide them to their destination. Satnav systems are gradually improving via the addition of information about congestion and alternative routing. Such information has been available for some time via dedicated sensors, but more recently some satnav systems have started to report on their speed of movement to a central control location. This can then make deductions about congestion (if many cars on a particular road report a slow speed, this is a strong indication of congestion), which it can then send to other satnav devices in the vicinity, allowing them to route around the congestion. This is simply achieved using cellular data channels and adds significant value to some drivers. It does require each satnav to have a cellular subscription, or alternatively to make use of a nearby cellular device such as the driver's mobile via a wireless link such as Bluetooth.

Such an application is making use of information collected by a large number of satnav systems as they move about in daily operation. As was discussed earlier, there are many applications of this sort where mobiles gather data passively as they are carried about and then send this data to a central unit able to compile local or national level information.

In principle, satnav devices could link into mobile-phone calendars, planning routes in advance and suggesting multi-modal travel arrangements (where more than one type of transport is used during a journey). So, for example, the car satnav could consult the driver's calendar, determine that the next destination was Heathrow and automatically load this into the destination address. It could go further and recommend that the driver stop at a train station to complete the last part of the journey by train.

In practice, this appears too complex for the rather limited gains that might ensue. Forming a link from the satnav to the home hub or driver's mobile could be problematic, depending on how close the car could park to the home. Setting up the system to allow for multiple cars and multiple drivers in the home would be challenging. Deducing destination information from details in the calendar could be very difficult (for example, the trip mentioned above might say 'Amsterdam' in the calendar, not 'Heathrow'). As satnav devices add voice activation, entering the destination will become increasingly simple, making the effort and complexity of automatically deducing the destination unnecessary.

The ability to auto-route across multi-modal transport possibilities is one that would appear to provide major benefits. A user would just provide the destination address and the best route would be deduced. If there were some problem during the journey, such as a train cancellation, then the navigation device could automatically re-route the user, taking all relevant information into account.

In practice, to do this would be extremely complicated. Very many user preferences would need to be understood, such as their trade-off of time versus cost and their preference for a longer journey time with fewer changes. These preferences might differ depending on whether the user was travelling on business or pleasure, whether they had luggage, whether they had family travelling with them, whether they had urgent appointments following the journey and so on. Even if all this could be accurately deduced, the difficulty of discovering all relevant transport information is great, since each mode of transport typically has one or more travel advisory services, different databases and different ways of storing and accessing the information. There would appear to be a very high risk of the navigation system getting it wrong.

Perhaps, then, the furthest that a car satnav might go is to look for parking at the end destination. To do this it might consult relevant web links embedded in its database of car parks, provide users with information on the cost and offer the possibility of advance booking where this might save money or be necessary.

Safety

There are many safety-related applications using wireless that have been suggested for the car. Closest to implementation is an initiative called 'e-call'. This places an automatic call to the emergency services if the car is detected to have been in an accident (for example, if the airbags inflate). The call would provide the location of the car and perhaps some information on its status (e.g. whether it is the right way up), allowing a more rapid response than would otherwise be the case.

Technically this is simple, requiring a cellular data link from the car and a small amount of software within the car. The problems are commercial. Such a data link requires each car to have a cellular subscription. In the case of many cellular systems this will mean that the car requires a SIM card. Having many more SIM cards in the network could be problematic to the operator since, even if these are not making calls, they will perform location updates as the car moves, loading the network. Hence the operator might expect some form of compensation. Another issue, as is so often the case with transport (and health), is that the benefits do not all fall where the costs lie. The benefits for e-call accrue to the vehicle occupants who receive help more quickly, but also to the local government, which is able to clear accidents rapidly, and, indeed, all other road users, who may experience less congestion. The costs meanwhile fall on the purchaser of the car and the mobile operator.

E-call is difficult for one vehicle manufacturer to introduce in isolation because agreements must be reached with cellular operators in all the countries where they sell cars and with the emergency services as to the number the system should call and the manner in which it would provide information. For this reason, it is better introduced by national or international governments and, indeed, e-call is an initiative from the European Commission (EC) which might lead to mandated inclusion of e-call systems in all cars sold in the EC. As part of this mandate, agreements with cellular operators and emergency services would need to be delivered. Predicting the future of e-call, then, is not about understanding the technology and economics but about understanding the politics of mandating new services at multi-national level. That makes it very hard to predict!

Another service that is often discussed is car-to-car (C2C) communications. Here one car sends information to others around it. The key application is emergency braking, where a car whose brakes are heavily applied sends a message to warn those behind to brake as well. Other messages could be envisaged, such as a warning of a slippery road if anti-lock brakes or traction systems engage. These concepts are worthy and clearly could improve safety – although they might not prevent the initial accident, they might prevent other vehicles from colliding with the car which has crashed. Technologically, they would appear relatively simple. A standard and set of protocols would need to be defined and a car would then automatically transmit the appropriate message. Cars in the vicinity would receive the message and take appropriate action. Spectrum at about 5.9 GHz has been suggested for a pan-European harmonised system.

However, there may be other problems with such a system. If it resulted in the brakes of all the cars in the vicinity being applied, this could result in cars going in opposite directions suddenly coming to a halt, potentially causing further problems. If it just sought to advise the driver of heavy braking then this could be done more easily, quickly and intuitively with brake lights that glow brighter under heavier braking. Other alternatives exist, such as forward-pointing radars in cars that automatically brake a vehicle if it gets too close to the car in front. These would work even if the car in front braked heavily and would not suffer the problem of affecting the wrong vehicles, such as those in other lanes. As forward-pointing radars become more prevalent for adaptive cruise-control systems, more cars will have the technology needed to stop themselves if the distance to the car in front decreases alarmingly rapidly.

At least a C2C system is simpler to introduce than e-call. It does not require anyone other than the car manufacturers to agree on standards and protocols, since the transmissions would be direct rather than via a cellular network. Even so, there is a 'chicken and egg' problem in that it is of little value for a consumer to pay for a vehicle with C2C emergency braking when there are no other vehicles on the road either to send signals or receive them. Only when most vehicles have the feature will it be of benefit. As car manufacturers strive to keep vehicle costs low, their incentive to put such devices into cars is very limited.

Even less likely is roadside-to-vehicle (R2V) communications. These could be, for example, signs that also transmitted wireless messages to the car to instruct it as to speed limits or warn of upcoming hazards. Or such a system could communicate imminent changes of traffic lights. Again, there are clearly potential benefits to these systems, but the benefits are relatively low. Similar information could be delivered by the satnav on the basis of its location and a database of known hazards and speed limits. Building such a system would require expensive infrastructure deployment, new technology in cars and standardisation across many countries and manufacturers. While this is technically feasible, it is very hard to see all parties coming together to invest and make this happen given the relatively low benefits.

Vehicle telematics

Many have talked about vehicles that can self-diagnose problems and transmit these to a garage, which can get the necessary spare parts ready and immediately resolve the problem. Again, while it sounds advantageous, it is hard to see clear benefits. Cars already have significant on-board diagnostics. Some of this information is displayed directly to the driver; other parts are retained within the engine-management system to be downloaded at the garage. Having the garage know the problems in advance would seem to have very limited benefits other than being able to ensure that adequate spare parts were available. In any case, it is relatively easy to transmit such information as long as a cellular data link exists within the car – the issues with making this available were discussed earlier.

Overall

There is much that wireless communications could do to enhance car journeys. However, many of these enhancements bring relatively minor benefits while requiring large-scale international collaboration across car manufacturers, local authorities and cellular operators. This explains why few of them have been introduced to date and it seems unlikely that many more will be introduced in the future.

Much more likely are wireless solutions that can be installed by the driver or by the vehicle manufacturer in isolation (i.e. without needing to work with other vehicle manufacturers and highway authorities). Satnav systems are an obvious example of this and their role is likely to expand to include reporting on congestion they experience and being able to route around congestion ahead. They might also provide warning information, speed-limit data and rudimentary journey facilities such as car-park booking. Satnav devices also have an advantage in that most are not built into the car but fitted as 'add ons'.

This allows them to be changed and updated several times during the lifetime of the car. With cars now lasting 10–15 years, this long lifetime can be a serious impediment to the introduction of electronic equipment that might prove to be outdated well before the car is scrapped. Other systems that can be introduced autonomously are forward-pointing radars linked to braking systems and simple solutions such as brake lights that glow more brightly as braking becomes heavier.

17.3 Rail

Railway applications can be divided into

- operational systems for the railways and
- passenger information systems.

Operational systems broadly help the railways to run trains more efficiently. They can do this by preventing problems happening, helping resolve them quickly when they do occur and enabling a higher density of trains on the lines.

Most possible railway solutions are based on a mix of location and tagging. One of the most talked about and probably most advantageous is electronic train control. Instead of controlling trains using static line-side signalling on the basis of their approximate positions, electronic train-control systems would receive a constant update of the exact location of each train. They would then adjust the speed of each train such that it maintained the minimum safe distance from the train in front. This would allow trains to be packed more densely and also offers potential savings by removing the line-side signalling systems, which are expensive to maintain and a source of frequent failures.

From a wireless point of view, as is so often the case, this is very simple. Location can be provided by GPS (except in tunnels, where other mechanisms may need to be employed) and the position can be reported using standard cellular technology. Control information can be returned to the train via the same cellular channels. The problems in introducing such a system are concerns about safety, the longevity of train stock (making retro-fitting difficult) and the costs of deploying appropriate wireless solutions. Self-provided cellular systems are often thought necessary by the railways because of a concern that public cellular systems might not provide the very high reliability that they require. Electronic train control has been discussed for decades, but implementation is rare and proceeding very slowly.

Other railway operation systems are based on sensing and tagging. For example, axles on carriages sometimes fail. A good indication of this is an axle running ‘hot’. Temperature sensors mounted in the track at strategic points can detect hot axles and read tags mounted on the bottoms of carriages. They can then report the problem and carriage identity to a central control point, which can schedule maintenance on the carriage. This can be simply achieved using an RFID form of tagging with the sensor either wired or using cellular data transmission to return its information. Many other similar sensor applications can be envisaged, such as sensors on overhead wiring that detect trains placing undue pressure against the power line. These solutions are simpler to implement than electronic train control since they do not involve replacing existing systems or changing operational practices and hence do not require safety certification. They simply add to the knowledge of the railway company. They can also be implemented on a small scale and gradually built out making a decision to go ahead much simpler.

Passenger information systems both provide passengers with information about the journey and also allow passengers to access information, typically by providing a WiFi service inside the carriage. The key problem here is getting a backhaul connection from the carriage to the Internet. Many train operating companies have solved this with a solution that comprises a mix of possible channels, with the best one being selected at any particular moment. Channels might include the following.

- *Dedicated backhaul systems.* The systems might operate at 5 GHz using masts set up alongside the tracks. These are effective but expensive.
- *Satellite backhaul.* This works well in rural areas, but buying satellite capacity is typically expensive per bit.
- *Cellular.* Where available, cellular solutions such as 3G high-data-rate channels can be effective.

There can still be some problem areas such as communications within tunnels, but, even if communication is lost in tunnels, this typically will not be too problematic for someone browsing, who will experience a delay whilst in the tunnel but see the connection re-established on emerging from the tunnel.

17.4 Air and sea

Like railways, air and sea can be divided into operational and passenger communications.

For ships, operational communications needs are relatively limited and mostly met already. Location is important, but adequately provided via GPS with backup systems like Loran if needed. Communication to shore is available using HF or satellite systems. If needed, cargo tracking can be provided using RFID.

For aircraft there are many more possible applications. One often mentioned is passenger baggage tracking. This could be facilitated with a RFID chip embedded in each bag tag, which could be read at various points in the baggage-handling process, allowing reasonably good location of the bag. While much discussed and with benefits that have been shown to be substantial, this has not yet been introduced. Part of the problem is in the RFID technology. Ideally, RFID sensors would be printable, allowing baggage tags to be printed at the checkout and making them very cheap. While there is much research in this area, the technology for RFID printing is not yet sufficiently well advanced for large-scale deployment. Another part of the problem is the need to standardise systems worldwide if they are to be effective. This is complex and, even if it is achieved, the speed of roll-out will be slow. In the interim, manually scanning the barcode on the baggage tag provides many of the benefits, albeit with the cost of manual intervention.

Other services for aircraft might include the removal of much paperwork, replaced with electronic communications. Fully electronic boarding passes may be implemented, although there are difficulties with this, since each passenger would need a compatible electronic device to store the pass and 'display' it when needed. This could be done as an application on a mobile phone, but not all passengers may travel with their phone, batteries might go flat or users may struggle with the actions they need to perform in order to 'check in'. Electronic systems might be used to download cockpit data as well as information on catering requirements and so on.

Passenger communications are similar to railways. Most passengers require Internet access and have devices with WiFi capabilities. Planes and ships can provide WiFi hotspots, but need to backhaul these to the Internet. In the case of planes and ships most solutions are based on the use of satellites, since ground-based communications are often not possible. An antenna on top of the plane or ship sends the data to a satellite, which re-directs it to an appropriate ground station. This works well, although the cost per bit on satellites is relatively high. There are some concerns as to whether there will be sufficient satellite capacity if this application is widely used, but satellite operators are considering launching new capacity in anticipation of growing demand.

17.5 Difficulties in introducing new solutions

Much like in the healthcare sector, there are many potential wireless applications in the transport area that would bring significant benefits and do not require any new wireless technology or standards. With congestion increasing and concerns over global warming, there are strong drivers for introducing many of these applications and yet progress is very slow.

One of the major problems is the mostly international nature of transport. Planes and ships mostly travel internationally, trains cross from country to country and cars are often taken to different countries. Even if vehicles stayed in one country, manufacturers would strongly prefer to fit a single solution that works in all countries rather than having to produce country-specific versions (with the exception of a few countries such as the USA and China, which are large enough markets in their own right). This means that any solution fitted to the vehicle that has to communicate with other vehicles or local infrastructure must be standardised across multiple countries and manufacturers. This is a difficult problem, given the large number of players involved and their often divergent interests.

Another problem is the safety-critical nature of many transport functions. Any new approach to controlling the speed or direction of any kind of vehicle has to be shown beyond all doubt to be at least as safe as the solution it is replacing. There are also possible legal concerns – for example, if a car fitted with an anti-collision system is nevertheless involved in an accident, is the fault that of the driver or the manufacturer? In such cases manufacturers often prefer to take the option of not introducing a new system unless mandated to do so.

A further problem is the longevity of most vehicles compared with the life-cycle of wireless communications. Cars routinely last 15 years, trains 30 or more and ships and planes longer than this. Wireless devices typically have a lifetime of a few years and even wireless infrastructure may be in place for only 20 years. So, for example, fitting a car with an embedded 2G radio system may appear appropriate at the time of manufacture, but 2G networks may be decommissioned before the car is scrapped. If the system is integral to the car's safety this could be problematic.

For these reasons, fitting wireless solutions in the transport arena has proven difficult, with a few exceptions where users can fit the systems themselves (e.g. satnav devices) or manufacturers can fit systems without any coordination with others or reliance on other systems (e.g. collision-avoidance radars). None of the problems preventing the deployment of wireless solutions look like they will ease in the future. Coordination will remain difficult and safety will remain an over-riding concern.

17.6 Prognosis

Wireless potentially offers much to the transport sector, but the difficulties in implementing solutions are huge. Transport lags behind the commercial sector by many years – for example, European railways are now implementing a 2G-based radio system (GSM-R) when cellular operators are trialling 4G!

Many of the solutions envisaged by futurologists are entirely possible today, but are unlikely ever to be implemented. The difficulties in achieving international agreement are just too great and the benefits too small, especially since there are often alternatives that are nearly as good but can be implemented more simply. The car-to-car braking system is a classic example where implementing brake lights that glow more brightly as the brakes are engaged more brings almost all the benefits for a small fraction of the cost and difficulty.

Transport will mostly remain well behind other users of communications in implementation of technology. Perhaps the most interesting case is the advanced satnav system, which could act as a source of information as well as receiving information and is able to re-route in order to avoid congestion.

18

Entertainment

18.1 Defining entertainment

Entertainment can be defined in many ways and the boundaries are blurred. For example, some classify gardening as entertainment; others regard it as a job around the home. For the purposes of this chapter it is relevant only to consider those areas of entertainment where wireless communications might play a major role. These might be defined as

- reading,
- listening,
- watching and
- playing.

Entertainment is a major potential driver of communications. The average person in the UK spends about $4\frac{1}{2}$ hours a day consuming entertainment, compared with about an hour for transport (time spent undergoing healthcare is not so easy to define). Measured by use of wireless resources, entertainment would undoubtedly be by far the largest consumer of wireless today, in almost any way that it was measured. Unlike healthcare and transport, entertainment is an area that can undergo sudden change, as the emergence of social networking has demonstrated, and where new devices and technologies can be rapidly introduced, as the iPhone has shown. It would be fair to say that entertainment is at the cutting edge of wireless communications and likely to remain there.

One of the key drivers for change in the entertainment sector is the Internet. In terms of reading, much material is now available on-line, making newspapers less attractive. In terms of listening, files can be downloaded from the Internet and readily transferred to MP3 players. In terms of watching, much TV content is now available on-line and viewing habits are shifting as a result. In terms of playing, the Internet is enabling multiplayer games and a whole new variety of ideas.

Reading

Reading covers a range of source material, including books, magazines and newspapers. More recently, reading has come to include material on the Web such as blogs, reviews and many other varied categories. Reading habits change only slowly – those who have been buying a daily newspaper for years tend to continue doing so out of habit even if there is clear utility in reading on-line. But habits are changing. Newspaper sales are falling slowly and, coupled with falling advertising revenues as advertisers move to the Internet, this is leading to major problems for newspaper publishers, with titles closing and the future looking relatively bleak. Magazine sales, conversely, are relatively healthy, despite the fact that the cost of a magazine per hour of 'entertainment' gained from it is the highest of all entertainment sources. This is probably because magazines are focussed on areas of particular interest or hobbies where people are willing to pay comparatively more.

E-books are currently a topic of much interest. E-book readers have been available for many years, without much success, but the most recent generation is increasingly user-friendly with paper-like displays, long battery life and in some cases the ability to download books directly from the Internet. On the face of it, e-books have overwhelming advantages such as instant distribution, small form factor and the ability to store an entire library on a chip and in principle e-books should be cheaper, since there are no printing and distribution costs. Yet change remains very slow. This is partly to do with habit – people have been reading paper books for many years and see no reason for change. Some of it is down to enjoying the look and feel of a 'real' book and having a physical library to peruse and admire. Books with pictures or photographs do not translate well to some e-book readers, particularly those with black-and-white screens. Also, publishers are wary of pirate copies and so are moving slowly and charging relatively high prices – sometimes higher than the print version.

Not much of this has any great significance for wireless. It might be that increasingly reading material is downloaded wirelessly, but, for the most part, files are relatively small. For example, entire books are only a few hundreds of kbytes and can take days or weeks to read.

Listening

Listening involves radio and music. These are both areas undergoing change. Radio can now be heard using analogue FM broadcasts, digital (DAB) broadcasts, podcasts downloaded to MP3 players and Internet radio, or through digital TVs. Much listening still remains on analogue radio. This is well suited to use in the car, as background when about the house and as a portable source of news and entertainment on the move. DAB is not progressing so well, despite the fact that many DAB receivers have been sold. Most homes in the UK retain analogue reception for the car and many other places in the home. DAB is not so well suited to reception on the move because battery life in portable units is much lower than for analogue FM and coverage at the higher frequencies of DAB is less than that of FM. With advertising revenues falling across all radio listening, the cost of providing DAB services is being questioned by many commercial providers. For wireless, radio broadcasting requires spectrum, but is well established and the amount of spectrum needed is relatively small compared with many other applications. Podcasting and Internet radio do not require wireless, other than perhaps a short WiFi link within the home.

Music is also changing, with more music being purchased electronically and downloaded to MP3 players. This does not generally involve wireless, since purchasing is typically done from a home PC via a broadband connection, although increasingly cellular operators are offering music download services direct to a handset. The size of a music file for an album can be greater than for a book, but is still relatively small compared with other data uses.

Watching

Watching involves video content. This is historically the area to which people have devoted most of their entertainment time and one that is undergoing much change. Video content can be received

- via over-the-air broadcasts through terrestrial TV (at 500–800 MHz) or satellite TV (at 12 GHz),
- via cables, through cable TV or DSL links over copper, or
- through watching a DVD or similar.

As well as these delivery mechanisms, there are the different ways of interaction that the user selects, which might be divided into

- passive, where the user just watches whatever is being broadcast at that point, selecting from the channels available on their broadcast service;
- active, where the user selects programming using some form of search, perhaps via an Internet-based TV service (such as the BBC's iPlayer); and
- time-shifted, where the user watches something they have recorded earlier, increasingly on a personal video recorder (PVR, sometimes known as a hard-disc recorder).

These mechanisms are changing and blurring – for example, the BBC's Project Canvas is developing a standard to allow Internet-based content to be selected and viewed on a TV (rather than on a PC).

The future of viewing is far from clear. Some suggest that viewers will increasingly move to active selection of content over the Internet and that the idea of a schedule and even discrete channels will fade. Others believe that the schedule still plays an important part and that there are many programmes that viewers want to watch live, particularly where they interact with them (e.g. in order to vote). Watching live allows viewers to share the experience with friends even if they are not in the same house by texting or otherwise communicating or more simply by talking together afterwards.

Another trend in broadcasting is towards higher-quality content. High definition (HD) is now available on many platforms, including satellite and cable, with the possibility of a limited number of channels on terrestrial broadcasting around 2011. As TV sets become larger, the shortcomings of standard definition (SD) become clearer to viewers, making HD more desirable. There is discussion around 3D TV and around even-higher-definition formats than HD, although it is not clear yet whether these will succeed. Over the coming years we can expect an increasing amount of TV to be delivered in HD format.

Trends in broadcasting can be very important for wireless. Broadcast content requires high data rates, especially for full-screen HD. If this is delivered via wireless, it consumes substantial amounts of the radio spectrum. Broadcasting TV in the UHF band already consumes about as much spectrum as all cellular communications systems combined (although this will reduce after analogue switch-off). In particular, if TV programmes are transmitted wirelessly to individuals, rather than being broadcast to all, then spectrum requirements will rise dramatically. In practice, much delivery will be via cable and DSL, but some will be wireless. In addition, there may be a growing need for wireless TV distribution around the home as TVs connect to the Internet and viewing moves towards active selection of programmes.

Playing

Playing predominantly involves computer games. Although they were for many years a minority occupation, devices like the Nintendo Wii have broadened the appeal considerably. While most game playing is not on-line, this is slowly changing and more interactivity can be expected in the future as players compete against others in different locations. Some have noted that they use interactive games not only as a means to compete with friends but also to communicate with them at the same time, using a chat function available within the game.

Gaming has not historically placed much demand on wireless networks. Data volumes have been relatively low, involving the sending only of data such as the position of a player, and game playing has typically been done via a PC connected to a home wired broadband connection. This may change over time as gaming on devices such as the iPhone becomes more popular, potentially leading to a large increase in wireless data related to gaming. Gaming typically requires data transmissions to occur with very little delay (so that a sudden movement made by one player is immediately seen by another) and this can be problematic for wireless systems, which tend to have relatively long delays due to the signalling and scheduling that needs to take place during data transmission.

18.2 In the home

Much entertainment is consumed in the home. Historically, the home was the only place to watch and play and one of the prime places to listen. This continues, although it is becoming easier to conduct all of these activities when out of the home. Nevertheless, many prefer to watch at home, where they have a larger screen and are not distracted by background noise. Similarly, playing is easier with a full user interface and a lack of distractions.

The home divides up into the broadband connection which brings data into the home, the home distribution system, which makes the data accessible around the home, and devices used for entertainment throughout the home. Each of these can be selected independently and in each home there could be a wide divergence of solutions.

The broadband connection

Most broadband connections to the home are wired, using cable, DSL, or in some cases, fibre optics. Wired connections are advantageous in that they offer high data rates and, in most cases, low contention.

Over many years there have been attempts to provide the broadband connection wirelessly, either as a competitive alternative in areas where broadband is already available, or as a means to provide rural broadband connectivity. Mostly these have failed, often resulting in the failure of the companies deploying the service and sometimes of those making the technology. Many different approaches have been tried, ranging from adapted cellular solutions through to bespoke 'fixed wireless access' and including advanced technology such as mesh. More recently, operators have been attempting to provide service using WiMax and there is discussion as to whether 4G technologies might provide an alternative broadband connection to the home.

For the most part, wireless has failed because it has not been able to provide sufficient data rates and capacity cost-effectively. As mentioned before, wireless capacity is determined by spectrum allocation, technology efficiency and number of cells. Providing fixed service does not materially change any of these apart from, perhaps, allowing higher frequencies to be used if an external receiver is mounted on the side of the house, in turn enabling access to more spectrum. This was the approach adopted by those developing mesh solutions, who accessed much larger spectrum allocations at 20 GHz or above as a result of deploying systems requiring only line of sight. Wireless technology is slowly becoming more efficient, especially at data transfer, but this change is not sufficiently fast to offset the rapid growth in home data usage, especially with Internet TV becoming more prevalent.

Wireless might play a role in two situations:

- where there is no broadband connection and
- where the home data usage is relatively low.

Most developed countries have extensive broadband networks. Only those living in deeply rural areas are unconnected because of the high cost of deploying a cabled system over long distances. Rural coverage also causes problems for wireless networks, since the number of homes within the coverage of a base station is low, especially if higher frequencies are to be used. This requires the deployment of a relatively large number of base stations, each of which needs backhaul. While the costs of providing wireless coverage may be less than for installing a wired system, they still tend to be higher than rural users are prepared to pay. Only with government subsidy are wireless solutions likely to be viable in rural areas. If subsidies are provided, then the use of cellular solutions enables the greatest economies of scale both for network and for subscriber equipment. However, such deployments will be of relatively small scale compared with national cellular networks.

Cellular systems have recently started offering datacards with tariffs of less than $20/month for limited amounts of data such as 3 GBytes. These are attractive for some households – often those where there is a single occupant who is renting and does not wish to set up the contract needed for a fixed line. But this is a very low-functionality solution. Typically, the connection is only to a computer or smartphone with 3G connectivity (although it is possible to connect a WiFi hub to these devices to distribute signals around the home). Coverage of high-data-rate cellular within homes is often poor, either preventing the use of data or resulting in low data rates. The cap of a few GBytes/month is adequate for web browsing and email but is soon exhausted if any kind of video is downloaded or streamed to the home. The obvious solution to this would be a higher cap, but this would cause problems for cellular networks. Already, high data usage on wireless cards is causing congestion problems on cellular networks and more widespread use of wireless as an alternative home broadband connection would require further capacity to be added. This is expensive and some simple calculations (see Chapter 2) show that at current price rates it is not cost-effective for cellular operators to enhance their networks. Essentially, they are offering broadband data at below full costs at present and thus cannot afford to expand the offering. By comparison, fixed-line providers are frequently increasing the data rates they can provide, lowering their prices and do not generally have caps on data usage.

With the exception of a few low-usage households and subsidised rural areas, wireless does not have a role to play in broadband provision to the home. It has failed to deliver fixed wireless solutions in the past and, with home data usage growing rapidly and ever-better fixed-line offerings, it cannot compete in the future.

The home distribution

The home broadband connection typically arrives at a single entry point into the home. It then needs distribution to all devices around the home. There are many alternatives to this, falling into wired and wireless solutions. Wired solutions typically make use of existing home wiring, since it is generally too difficult to install new wiring in existing homes. This includes coaxial distribution systems where fitted or using the home power lines. Both work well.

Wireless solutions are popular both because they are cheap and simple and because they allow portable devices to be used throughout the home – such as taking a laptop to whichever room is preferred. WiFi is well established as the home distribution system and is being enhanced for ever higher data rates.

As mentioned when discussing health, there may be some cases where hybrid solutions are appropriate, employing shorter-range wireless solutions covering part of a house. These might be to link to sensors for which it is not appropriate to use WiFi because it may drain their batteries. Alternatively, higher-data-rate solutions may be needed. For example, streaming of video from a DVD player to a screen typically cannot be performed using WiFi because the data rates are too high. Alternative solutions include ultra-wideband (UWB) and links at higher frequency bands such as 60 GHz. These typically have sufficient range to cover only a single room. As a result, a means of distributing data to a point within each room from where it can be radiated within the room, is required. If the short-range system operates at a high data rate the home distribution will probably need to be via wiring so that it can feed each of the rooms in the house.

There may be multiple systems within the same home. It might prove simpler for different applications to have their own hub – for example, there might be one for the entertainment systems and another for the smart grid or home energy solution. Separating the hubs makes it simpler to install new systems like home energy or home healthcare because the hub and devices can be supplied preconfigured and need to take little account of what is already available in the home. The disadvantages of multiple hubs include possible interference and difficulty in sharing information.

Regarding interference, if all the hubs are wireless and are using the same frequency bands, such as the 2.4-GHz unlicensed band, then, unless they are correctly configured, there is the possibility of interference. Also the greater use of the band will increase overall levels of congestion, which might be problematic where homes are in close proximity and their systems cause interference with each other. In this situation, whereas beforehand each of, say, four homes close to each other might have used a separate channel in the 2.4-GHz spectrum (where there are four discrete separate channels when using WiFi systems with a bandwidth of 20 MHz), with multiple home hubs each home might now use two channels, resulting in neighbouring homes being unable to find clear channels and interference occurring.

Regarding the sharing of information, all hubs would need to connect to the broadband router, but this would typically not be able to sensibly route information from one hub to another. In many cases this would not be an issue – for example, it is hard to envisage circumstances in which the home entertainment system might need to interact with the home energy-management system. Where it is an issue, it might be possible for hubs to effectively communicate through the Internet – both accessing a site in the cloud where they can share information in a pre-defined manner.

Devices in the home

The purpose of the broadband connection and home distribution system is to enable devices in the home to communicate, either with the outside world or with each other. Devices in the home are already varied and this variation can be expected to increase in future. Home devices include, for example,

- computing systems, including desktop and laptop PCs;
- gaming systems, including those linked to a TV screen, such as the Nintendo Wii, and those with their own screen, such as the Sony PSP;
- communication systems, such as phones and intercoms;
- TV sets, including associated DVD players and PVRs;

- home sensors, such as security cameras and health-related sensors;
- portable devices, including cameras and calendar devices; and
- mobile phones.

With all of these, and more to come, each home is likely to be unique in terms of the devices it has and how it wishes to use them. Further complexity arises from most homes having multiple occupants, all of whom might wish to use the broadband connection simultaneously and for different uses. This complexity could become a major issue in the future, with home networks requiring careful management and occasional attention.

One of the key architectural questions for home devices is whether they will be autonomous, simply using the home hub as a means to connect to servers and services outside the home, or whether they will be coordinated by the home hub which will act as a 'server' for them.

Autonomous devices are much simpler to design and to use in the home. Each device simply needs to connect to the home system, e.g. via WiFi, and can then immediately be used. No standardisation is needed and very little, if any, configuration needs to be performed by the user.

Coordinated devices need to be configured into the home system. The home server can then interwork with them, passing them information or retrieving it as needed. This allows, for example, any TV in the home to access content on the home server or potentially in any other location, rather than just retrieving content from the Internet or broadcast services. It allows sensor systems to be controlled from the home PC, with action being taken directly if needed. It enables relatively simple synchronisation of video and audio content across a range of devices, for example automatically downloading the latest music to a mobile device when it enters the home.

Clearly coordination brings benefits, but it also brings complexities and the need for different standards to enable appropriate interworking. For example, TVs would need to know how to request a list of available programmes from the home server and how to stream these when one is selected.

A compromise might be for all content to be hosted outside the home, in the cloud. So, for example, a home owner might upload their videos and music to a server-based library. Devices in the home would then access this library rather than trying to download files from a local computer. This might simplify the configuration task, since there is less variability in linking to a web site with content on it than there is in linking to a PC or server in the home. It also allows content to be accessed when outside the home, for example when mobile or when visiting the homes of others. It has the disadvantage that bandwidth will be consumed across the broadband connection and that congestion in the core network or the server site might cause delays in downloading content, but these disadvantages can probably be managed such that they do not become significant.

The likely outcome, as is so often the case, is that all options will exist, in some cases in the same home. Some devices, such as gaming platforms, will be autonomous since the benefits from connectivity are low. Some TVs in the house may be connected to the home server, with secondary sets perhaps working autonomously. Some homes will have dedicated home servers; others will use the primary PC to coordinate activities. Some standards will emerge, such as the Digital Living Network Alliance (DLNA) standard for sharing content (predominantly video, music and pictures at present) throughout the home.

18.3 On the go

Entertainment outside the home tends to be restricted by the devices that can be carried and the battery life available. This tends to make entertainment on the move a subset of entertainment in the home. Watching is possible, but only on small screens. Listening is mostly through headphones. Reading is via either paper or e-books. Gaming is mostly standalone on portable game players.

Many of these devices do not need a wireless connection. Often the content is 'sideloaded' onto the device while in the home rather than being downloaded as needed. With device storage growing quickly, many devices can store far more content than the user could consume on a single trip.

A key uncertainty is whether this will change in the future. Content may be stored centrally in 'the cloud' or on home servers and accessed when needed. Or users may pull down new content whenever they wish rather than waiting to return to the home. Accessing content on the go is clearly important for live content such as news, sporting events and weather. Such dynamic access to content could have implications for wireless networks.

The key question for wireless networks is whether they will be able to transmit the potentially very high data loads that might result, especially if video content is routinely downloaded to multiple devices. The extent of any problem is almost impossible to predict because it depends on factors such as

- the amount of video the average user downloads;
- whether they download and watch in a format compressed for a small screen, which results in much lower data rates; and
- the type of network to which their device is connected when they download, for example waiting until within the coverage of WiFi for large downloads would hugely ease the loading on cellular networks.

Models have shown clearly that using scenarios that generate the highest amount of traffic on the cellular network will lead to congestion even assuming ever greater efficiency of cellular technologies and the provision of substantially more spectrum for cellular applications (as was discussed earlier in Chapter 2). Some have taken this to show that efforts to expand cellular capacity need to be redoubled, but there is an alternative interpretation – that these scenarios will not come to pass because pricing on the cellular networks will be such that they will be discouraged.

The most likely outcome is that capacity on cellular networks will remain relatively constrained and expensive. Much entertainment content will be side loaded to avoid these costs. Dynamic downloading of content will be restricted to material that is live and has value to the user and, even then, the always-best-connected device will seek to download this content over low-cost networks such as WiFi. Entertainment will expand to fill the network available rather than the network expanding to provide all possible requirements.

18.4 Key trends

Entertainment appears to be an ever-growing and rapidly changing sector, although it is worth remembering that most people already devote most of their free time to entertainment so it is unlikely that the time spent will grow, but the way the time is spent may change. Many of the changes will affect the entertainment value chain, with newspapers suffering while Internet-based services grow. These changes have not been discussed in detail here because they do not materially affect the architecture or level of use of wireless networks.

The amount of information being transferred as part of entertainment appears to be growing quickly. Although reading and listening do not result in particularly large files, pictures and particularly video can generate massive download requirements. Trends towards higher-definition video and personal download rather than broadcasting a programme simultaneously to many viewers mean that these requirements could grow substantially.

Perhaps counter-intuitively, these trends are likely to result in less use of wireless to the home. This is because wireless networks have always been second-best to wired networks and their capacity is growing less quickly than data requirements, making wireless increasingly less attractive than a high-speed wired connection.

However, there is an ever-increasing trend for more wireless within the home and perhaps for more short-range wireless in key locations in the environment to relieve capacity pressure on cellular networks. This wireless is likely to be provided using WiFi solutions because WiFi is ubiquitous, inexpensive and provides high enough data rates for most applications. Conversely, the wiring in most homes is inadequate for high-speed distribution and is likely to become more so as needs rise. Only in new-build homes might adequate wiring solutions be installed, but many decades will pass before a large percentage of homes can be equipped.

The greater use of wireless in homes and hotspots raises an issue mentioned before – the possibility of congestion and interference. Congestion might occur as more networks are deployed and used more intensively. Any congestion or interference would be increasingly problematic because users rely more on WiFi as part of their everyday life. Already failure of cellular networks is problematic to many and failure of WiFi could become similar.

19

The smart grid

19.1 Introduction

At present, most electricity networks simply supply to the home or office whatever energy is demanded. A meter, often at the periphery of the building, monitors consumption and the building owner is subsequently charged accordingly.

While simple, this approach has a number of disadvantages.

- It requires significant additional electricity generation to be available in order to supply peaks in demand. This is both costly and can have an environmental impact.
- It provides little information to the home owner as to their instantaneous usage, making it hard to understand how energy consumption can be reduced.
- It does not readily allow electricity generated locally, for example via solar panels on the building, to be supplied back to the grid.
- Reading the meter can require the visit of an employee to the home.

With increasing environmental concerns and the possibility of a substantial increase of demand for electricity if battery-powered cars are charged at home, there are strong drivers to enhance the electricity supply in order to overcome these disadvantages. Such an approach is often termed the 'smart grid' – 'smart' because it would have some intelligence in terms of the way in which electricity is consumed. There are many differing views as to what the smart grid might look like and how it might be provided, which are explored in this section.

19.2 Architectures and connectedness

There are two elements of a smart grid – either or both may be present in any implementation. The first is a communications link between the home and the electricity supplier. This enables features such as

- the supplier signalling changing prices to the home as demand changes;
- the supplier exercising more direct control over usage in the home, for example causing less important appliances to pause during peak demand;
- direct measurement of consumption, avoiding the need for manual meter reading; and
- the supplier being able to monitor the voltage and any distortion in the electricity signal, which allows them to better tailor the signal being delivered to sub-stations.

The second is a link within the home between elements such as the meter and major appliances and a home control point. This enables features such as

- provision of overall consumption information to the home owner such that they are aware of the costs of their usage,
- provision of consumption information from individual appliances and
- scheduling of appliance operation to minimise energy costs.

There is some overlap between these – for example, the supplier cannot exercise direct control over home usage without links within the home to the appliances.

There are alternative architectures that can be used to realise these elements. Information from the supplier could either be carried over a dedicated network – perhaps over the powerlines themselves using powerline telecommunications (PLT) – or provided over the telecommunications link to the home – perhaps via a site on the Internet that is regularly checked by the home energy-management system. Where information is provided by the electricity supplier, this tends to result in the home meter becoming a key element that decodes the supplier's signalling and signals into the home accordingly. Where information is provided over telecommunications channels, the meter may have limited functionality – being there primarily to record usage – but there is a need for a 'home server' that manages appliances throughout the home. The former allows the electricity supplier to be very much in control of the smart grid; the latter limits their role to the provision of pricing or demand information. In either case, the amount of information exchanged is relatively small, but there are important implications for the evolution of home networks, which will be considered in the next section.

19.3 Home networking and the smart grid

The choice between a dedicated smart grid and one hosted over the telecommunications network depends in part on the evolution of home networks.

If a hosted smart grid is to fulfil its potential, it requires a home network with a home server, probably always on, connected to the Internet and networked within the home using one of the home network technologies discussed in earlier chapters such as WiFi or home PLT. Such a home server is unlikely to be implemented solely to deliver a smart grid, but may be part of a broader home-entertainment and -management system. Were such home networks to develop then it would appear more logical for the smart grid to be integrated within them. This might allow, for example, both the control of when appliances operate, in order to avoid peak demands, and also maintenance, control and updating of software. Communications to a battery-powered car, for example, might not only control the time at which it was charged but also update the satnav system and provide driving information to insurance companies. For such a hosted smart grid there is little need for infrastructure development outside the home, but substantial investment within the home would need to be made by the home owner, probably as they replace home PCs and entertainment systems.

Conversely, a dedicated smart grid requires very little investment within the home. Each major appliance could be provided with an insert that sat between the wall socket and the electricity plug and was controlled by a smart meter using PLT communications. However, there would need to be substantial investment from the electricity supplier in enabling PLT over their supply network and in upgrading home meters. The approach of simply turning the power to appliances on or off also limits the number of appliances that can be controlled. For example, if power is turned off to most washing machines, they do not restart automatically when it is resumed but require a manual restart of the washing programme. While fridges and freezers will restart automatically, some may generate 'power outage' warnings, which could be annoying for end users. Further evolution of appliances such that they can accommodate better having their power cut and restored might be needed.

The risk in adopting a hosted smart grid is that insufficient homes will upgrade their home networks to realise the benefits of load smoothing. The risk of a dedicated smart grid is wasted investment by the utilities and a lack of flexibility in the home due to separate control networks. There may also be problems with obsolescence in that meters are currently expected to have a lifetime in excess of 20 years. During this time, communications protocols and standards might be expected to change significantly. For example, if the meter used Bluetooth to signal to a display panel within the home, it is possible that within 10 years or so the Bluetooth standard might have evolved or been superseded and that chipsets for the old standard would no longer be available, making it impossible to deliver replacement home displays. There are possible solutions, such as 'plug-in' communications modules and long-term supply contracts, but these will require careful thought.

The two types of network can exist in parallel, with home owners selecting between them, and changing from one to another over time if they so wish. This may be the most pragmatic solution and is perhaps the only workable one where funding for the smart grid is provided by the home owners and utilities. However, where there is government funding to enable the smart grid, there may be a desire for more centralised control and more rapid roll-out in order that the benefits of reducing electricity-generation capacity can be more rapidly realised.

The solution selected also influences the manner in which utilities control demand. With a dedicated grid, the utility can offer a lower tariff in return for its ability to directly control demand within the home. With a hosted grid, the utility cannot exert direct control over the home but instead needs to signal to the home network, probably via spot pricing, whether it would like to see a reduction in loading. The home network can then act upon the pricing information in a manner determined by the user (or using default values provided by the utility). Implementing spot pricing might not be appropriate for homes that do not have any means to control their appliances, but conventional tariffs could be maintained alongside spot pricing. The potential savings from spot pricing might be the inducement to encourage home owners to deploy their own network.

19.4 The need for a compelling home networking application

Clearly the most logical solution is the hosted grid. This provides much greater flexibility, builds on existing communications networks and allows integration of electricity demand with other aspects of control within the home. However, there may be many reasons why the most logical approach does not prevail, including the behaviour of the utility, concerns about speed of adoption of home networks and the question of government subsidies. This is more of a social and political issue than a technical one and correspondingly difficult to predict at this point.

20

Assisted living

20.1 An ageing society

The statistics around the age and well-being of our societies are clear. There will be an ever-increasing percentage of older people whose healthcare needs will grow. At the same time, the number of people available to act as carers will fall, as will the number of people in work and paying taxes to pay for healthcare. Without changes to the manner in which we deliver healthcare, the situation will become increasingly problematic.

It is likely that there will be many elements to the solution. One of these may be the use of technology to allow people to stay in their homes for longer and to live healthier lives. Wireless may have a role to play as discussed in this chapter, which builds on earlier chapters looking at technology and wireless in the healthcare arena.

Assisted-living solutions can be divided into the following categories.

- Telehealth – delivering medical care, treatment, or monitoring services to old and disabled people at home from a remote location.
- Telecare – delivering social care/monitoring services to old and disabled people at home from a remote location. This includes preventative care.
- Healthy-living services – delivering services for healthier lifestyles to old and disabled people at home from a remote location.
- Engagement services – delivering services into the home from a remote location, to engage older and disabled people in terms of social, educational or entertainment activities.
- Teleworking – working remotely from home for an employer, or voluntary work, or as a self-employed person who needs remote computing to work successfully with others.

Many of these have similar elements – they are mostly about the delivery of a service to the home from a remote location. In some cases entertainment solutions can be used to help with assisted living – for example, gaming programmes can provide feedback on how alert and freely moving the individual is. In this chapter we consider the mechanics of delivering the service, the need for additional technology elements and the difficulties of implementing services of this sort.

20.2 Delivering assisted-living services

Delivery of assisted-living services does not generally differ from the delivery of other services into the home. There is a need for a service platform connected to the Internet, a data connection into the home and a mechanism for distributing and manipulating the data within the home. It seems unlikely that assisted-living services will place any greater requirement on home delivery than entertainment services and hence existing broadband and home distribution solutions, as discussed in earlier chapters, will be adequate.

One issue is that the elderly are, at present, less likely to have a broadband connection and home computer than the rest of the population. This is generally not because of lack of availability but because of lack of familiarity or a perceived lack of need for such a service. This is likely to change as those who currently do have broadband age and as government initiatives help, and, if assisted-living services showed clear benefits, then uptake of broadband would probably grow as a result in any case.

Within the home there will need to be a means of viewing information and possibly a sensor network distributed around the home, as discussed earlier in Chapter 13. Some types of sensor may be specific to assisted-living situations, but this will not change the overall architecture or issues of integration. There may also be a need for simplified PCs or bespoke assisted-living home platforms that make interaction simpler so that they can be used by those unfamiliar with technology or with disabilities.

20.3 Difficulties

The difficulties with assisted living are similar to those for many other sectors – less technical and more to do with the need to coordinate many different entities. Unsurprisingly, they share many of the problems of the healthcare sector, in particular a lack of funding and problems to do with the fact that the bodies which would provide the funding are different from those that would reap the benefits.

The benefits of assisted living are still unclear. Because few solutions have been tried and there are so many variables, it is difficult to be sure whether assisted living can make a significant difference to healthcare and to the lives of the elderly. Often with complex societal systems there can be unanticipated side effects – for example, allowing people to spend more time in their homes might lead to an increase in the number of falls or an increased cost in modifying homes to enable the disabled to live in them. Without large-scale trials, the evidence as to the benefits will not become clear. Equally, without the promise of large-scale deployments, appropriate sensors, software and solutions will not be developed or will remain expensive. Just as described with healthcare, a piecemeal approach may be simpler and quicker to implement, but in this case may be insufficient to enable people to stay in their homes. Some trials are now starting and may provide useful evidence in 2011 onwards, but equally the results may be inconclusive.

In many countries coordination between different parts of the healthcare services may also be insufficient. This may make it difficult for appropriate bodies to consult medical records or for those implementing the assisted-living services to work effectively with doctors and hospitals to ensure that appropriate telehealth information is provided and any information generated by the assisted living system is available in whichever parts of the healthcare system need to consult it.

There may also be ethical concerns over the intrusiveness of some solutions. This may make them less acceptable to patients and cause delay while their merits are debated.

20.4 Prognosis

The prognosis for assisted living is very similar to that for healthcare – both provide monitoring services in the home. Just as with healthcare, there is generally no need for technical advances to implement assisted-living solutions. Broadband to the home coupled with existing wireless distribution mechanisms within the home will be sufficient. However, the speed of implementation is likely to be slow because of the difficulty of introducing new technologies and ideas into large, complex healthcare providers and because the incentives for investment often do not align with the realisation of the benefits. Most likely is that assisted-living solutions will initially be provided by the individual, or other members of their family who see advantages in remote monitoring. If this proves successful then over time healthcare bodies may move to provide some elements themselves or assist some individuals in implementing their own systems with a mix of funding and advice.

21 Universal service

21.1 From a nicety to a necessity

A decade ago the mobile phone was still something of a novelty. Users accepted that coverage was imperfect and were mostly happy to be able to make calls in a few locations. Dropping calls when passing through areas without coverage was accepted as a fact of life when using mobile phones.

Over time people have become increasingly reliant on the mobile. Instead of a tool to allow plans to be adjusted at the last minute when necessary, it has become the means by which peoples' lives get organised. Tradesmen have dispensed with front-office staff because they are able to take calls while working. Businessmen arrange conference calls while travelling on the assumption that they will have mobile coverage. Parents leave children unattended, assuming that they can be contacted via a mobile phone in case of problems. A lack of mobile coverage can be a significant problem to many in their lives.

Areas where there is no coverage are often known as 'not-spots' and the complaints about these have increased steadily. An expectation of perfect coverage is growing – both from consumers, who want reliability from their phones, and from governments, which want to provide citizens with services that are increasingly seen as essential. This section looks at why not-spots occur, how they might be resolved and whether they will lead to a mobile service that is increasingly regulated and delivered as an essential utility.

21.2 Not-spots and not-zones

A not-spot is not a well-defined concept. Broadly, it is a location where a user cannot get mobile coverage. It may be a not-spot just for the operator to which the user subscribes or it might be a not-spot across multiple operators. A not-spot might be a few metres across, perhaps caused by the shadow of a building, or an area many tens of miles across (a 'not-zone'). Not-spots can also vary across mobile services. A complete not-spot would not have coverage for any kind of mobile service: voice, texting or data. A partial not-spot might allow voice but not data, or it might allow only low-speed data. Not-spots can vary according to the user. Taller users might get a stronger signal. Different phones have different qualities of antenna, so a not-spot for one model of phone might not be a not-spot for a different model of phone. Not-spots can be time-variant: coverage is often worse in the summer, when the trees have leaves and absorb more of the mobile signal, than in the winter. A significant complaint is coverage in the home or other buildings, where a not-spot might exist in part or all of the building despite there being coverage outside.

Not-spots, then, are very complicated and difficult to define. All would agree that a complete lack of coverage in a train tunnel is a not-spot, but a low data rate in the basement of a rural building may be acceptable. Complete coverage everywhere for voice and high-speed data, including in the basements of buildings, is unlikely to ever be possible, but a higher level of reliability might be achieved. In developed countries coverage for 2G voice is often delivered to about 98% of the population. Achieving 99% or even 99.5% might be possible.

Mobile operators, of course, would prefer that their network provides perfect coverage, but enhancing coverage costs money and operators also need to make a profit. Broadly, not-spots remain because it is not economic for the operator to fix them. To resolve a not-spot typically requires the deployment of a new base station, which has both a capital and an operational cost. To justify the cost, a certain volume of calls must be made in the cell, which will happen only if there are enough subscribers living there or passing through. Too few subscribers and the cell will not be economic and hence probably not deployed. In some cases, the not-zone might extend over large rural areas that might need tens or even hundreds of base stations, so, if few people live in the zone, the cost of coverage cannot be justified.

The economic argument is not quite as simple as considering a cell in isolation, because subscribers like to know that their operator provides coverage in remote locations in case they have an emergency in these areas. Here providing some coverage that is not directly economic might generate more subscribers in other areas, so paying for the cost of the cell. This is the reason why most operators provide coverage on major roads through remote areas.

The solutions to not-spots are those that make them economically viable, either by reducing the cost of providing coverage or by finding an alternative mechanism to pay for the coverage other than through call costs. There are many elements to the cost of a cell site, so there are many ways to make savings. For some time the cost of finding a site has been high due to major local objections to masts, requiring multiple rounds of planning applications and much legal and planning effort, often, perversely in those areas where not-spots are a major problem because these are often areas where there is the strongest desire to preserve the natural beauty. However, attitudes appear to be slowly changing, with consumers realising that, if they want mobile coverage, then they have to accept masts, and hence the costs of planning coverage are slowly falling. Base stations are also becoming less expensive, especially low-capacity units where femtocells and picocells can be used rather than conventional large equipment racks. As femtocells are produced in large quantities for the home markets the economies of scale can result in a step change in equipment cost for rural areas. Other areas of cost reduction such as lower-cost microwave back-haul can also have some effect. Finally, automated cell planning can reduce the cost of integrating new cells into the network and of updating various parameters that need periodic adjustment for the continued effective operation of the cell.

Network sharing is another possibility. Here operators share various elements of the infrastructure in order to reduce the cost. Many operators now share masts, placing their antennas on the same mast and their equipment at the foot of the mast. Some go a stage further and share the electronic equipment and the backhaul. However, most of the benefits of network sharing come from sharing the mast – of all possible savings, about 80% comes from mast sharing. Sharing the electronics brings fewer gains because, if one operator seeks to use the electronics of another, they will still typically need to install more cards and amplifiers to handle the extra calls, negating some of the savings. Sharing electronics also reduces flexibility, since both operators then typically need to upgrade at the same time and often adopt the same software releases, reducing their ability to compete with each other. A more extreme form of sharing is for one operator to cover a region of the country and a different operator to cover a different region and for their subscribers to roam from one network to another. However, this might not be allowed by regulators and can be difficult to implement.

If the operator cannot get the cost low enough then another solution is for subsidy to be found. One option is for a local community that wants coverage to pay some money towards the cost of a base station or to acquire and build a mast site that operators can use, perhaps at no cost. Or regional development agencies might provide funding to improve the coverage in their region. For example, this happened in Scotland in the mid 1990s when the regional governmental agency paid some money towards coverage being provided in the rural 'highlands and islands' area. This coverage was provided by both of the major operators of the time, on shared masts. Various methods of governmental funding of coverage are considered in the next section. At the most local level not-spots within a home can be resolved by the home owner acquiring a femtocell (as discussed in Chapter 3), whereby they are effectively paying for the entire cost of coverage – the base station, the installation, the backhaul and the electrical power.

There are a few special cases of coverage, particularly on planes, trains and ships. Coverage on planes and ships is broadly achieved with WiFi or femtocells within the plane or ship and satellite backhaul. Trains are more complex because they travel through areas where some coverage is already provided. In the longer term, it seems likely that coverage on trains will be achieved with repeaters or active units within the train and backhaul provided via a mix of satellite, cellular and bespoke wireless communications from the roof of the train. In the shorter term, better signal levels alongside the track are needed in order to get coverage within the train. Special solutions are needed for tunnels and deep cuttings where typically a 'leaky feeder'[1] needs to be deployed.

21.3 Regulatory intervention

As consumers have become more concerned about not-spots but levels of coverage have not changed materially, governments and regulators have been considering whether they should intervene. Intervention risks impacting on competition and can have unintended consequences but, since coverage is seen as increasingly important, some are concluding that it is worth taking these risks.

Intervention could take many different forms. We have already mentioned direct funding. Here central or regional government might invite operators to bid to provide coverage across certain regions. The operator making the lowest acceptable bid would get the funding. Typically the operator would not need to be subsidised to the full cost of the deployment since they would expect to get some revenue from consumers using the cells and so will often seek only something in the region of 10%–50% of the cost of the coverage provision. However, coverage from just one operator is of value only to those consumers who are subscribers with that operator unless some form of roaming is imposed

[1] A leaky feeder is a piece of coaxial cable with small holes created at regular intervals where the signal can 'leak' out. This allows a constant level of signal throughout the length of a tunnel.

(as discussed below) and so provides only a partial solution. Governments also generally do not like having to provide funding for commercial companies.

An alternative to directly funding coverage is to fund it indirectly through coverage obligations. These are requirements in the licence of an operator to cover a certain percentage of the country or population. It is a form of indirect funding because, if the coverage requirement is greater than that which the operator would normally have planned for (and there is little point in the obligation if it is not), then the operator will bid less for the licence, by approximately the amount they expect the extra coverage to cost. The government will then get less revenue and so will have effectively paid for coverage. Economically, this is an inferior approach insofar as it does not make the cost of coverage clear and obvious and so is less accountable. Politically, it is expedient since a politician can claim that not only did they get revenue from the auction of a spectrum licence but also they secured coverage 'for free' across their constituency. There are other problems with coverage obligations, such as exactly how to verify them and what sanctions to impose if they appear to have been breached, but these issues often seem some way off at the time of an auction.

A less interventionist approach is to ensure that consumers have access to good information. If consumers are able to discover exactly what level of coverage is provided by each of the competing operators in regions and along routes that are important to them, then they can make an informed decision regarding their choice of operator to which to subscribe. If one operator has materially better coverage in one region, then they might gain subscribers from their competitors, who will eventually notice and improve their own networks to stem the losses. In this manner good information can encourage better levels of coverage without any direct intervention. Simpler information on what consumers might reasonably expect by way of coverage in areas such as within their homes and how they might expect data rates to vary within their homes can also prevent frustration and complaints, while providing information on solutions such as femtocells can provide options for those who wish to act.

Providing good levels of coverage information is easier said than done. Many operators estimate their coverage using propagation modelling tools that use the positions of the base stations to predict signal levels. These tools have been developed and enhanced over the last 20 years to the extent that they provide generally good estimates over areas of several hundreds of metres but are less good at providing details of smaller not-spots. They also occasionally make larger errors. Operators often provide access via their websites to the predictions from these tools to subscribers, who can enter their home location and see a stylised coverage map showing the mobile services predicted to be available in the vicinity. A subscriber could visit the prediction websites for each of the competing operators and select the one with the best coverage in areas important to them. However, this is often difficult in practice because different simplifications to the data are made by different operators, making comparison difficult. Better would be a single site that provided a comparison across all operators on an equivalent basis and ideally using signal levels gathered through measurement rather than via prediction. Developing such a site would be costly, but not impractically so, and is currently being considered by a number of regulators.

A final option, mentioned earlier, is to require roaming between national operators. Roaming is typically implemented between international operators such that on arriving in a new country a subscriber is able to use one or more of the networks in that country, but is rarely implemented on a national level. For national operators there is little incentive to allow their subscribers to roam onto their competitors' networks and hence lose the call revenue. If governments require national roaming, they risk reducing the incentives for investment since there is little point in an operator providing better coverage in a region if they know that their competitors will then simply enable their subscribers to roam onto this site. Only if the investing operator can charge premium rates to roaming subscribers might deployment be worthwhile, but this then raises issues around variable call rates depending on the network the subscriber is on and means to signal this to the subscriber in order to allow them to control it.

The effectiveness of roaming depends on how many not-spots have coverage from at least one operator. If all not-spots are areas without coverage from any operator, then roaming will not bring any benefits, but if most not-spots have coverage from at least one network then roaming can do much to eliminate them. Which of these is generally the case is not clear, although, as network sharing is more widely implemented, the coverage patterns of those operators sharing masts will tend to become more similar, meaning that where at present there is just one operator providing coverage in a not-spot, there will be either no coverage at all or coverage by more than one operator.

At present, a few regulators have mandated national roaming, especially in rural areas, but most have not. Whether this will change remains to be seen.

21.4 Resilience and reliability

A related point is the resilience of networks. With greater reliance on mobile networks to summon help in emergencies, a failure of a mobile network can be life-threatening. Governments and the public are increasingly concerned that networks do not fail, particularly during times of high stress such as when there has been a terrorist attack in a major city. Keeping a network in service during such a time is problematic, not so much because the incident might have damaged the infrastructure (mobile networks are typically dispersed and most incidents affect at most a few base stations) but because the resulting peak in network usage places an unusually high load on the network. Designing networks to carry such loads is typically uneconomic because they occur so rarely, but networks can be designed to reject any traffic they cannot accommodate while still carrying traffic up to their maximum level. Most mobile networks now implement some form of 'access control' that allows them to regulate the demand such that the network itself does not fail.

The resilience of any given network can be understood only in the context of a number of specified scenarios. It is impossible to be sure that a network is resilient with respect to any future event – for example, terrorists could decide to target every base station in a network, against which threat few networks would be resilient. In some countries designing resilience against earthquakes is important; in others flooding is considered the major risk. Some problems are known to occur periodically, such as cables being accidentally cut, and hence networks are often designed to accommodate such events through diverse cable routing. Of course, specifying possible future scenarios is uncertain and hence it is never possible to be absolutely sure how resilient a network is.

There are some general design guidelines that are often followed to try to ensure that the network fails as little as possible. Main cable routes, such as from a switching centre, are typically duplicated with diverse routing such that, if one cable fails, the alternative can rapidly be used. Major points of failure such as key registers in the core network are often duplicated, or have duplicated elements within them on 'hot standby' so they can be quickly swapped in if needed. Backup power supplies can be provided both in the core and at the base stations, using batteries and generators.

Alongside all this best practice, over-dimensioning of the network capacity is often the best way to achieve resilience. In many problem situations the traffic loading increases. This may be because more people are trying to make calls, because another network has failed and traffic is being shifted (e.g. people using their mobile phones instead of the fixed line if that has failed) or because part of the network has failed (e.g. a base station) and nearby elements are trying to service the traffic. A network that has ample spare capacity will be able to provide service in these situations whereas one that is running near capacity will at best not be able to handle the additional traffic and at worse may suffer reduced performance due to being unable to adequately refuse the additional traffic. The latter is what happens during 'denial of service' attacks on websites.

Hence, a good proxy for the ability of a network to survive unknown future scenarios might be the level of over-capacity it provides.

Providing resilience either in terms of redundancy or in terms of over-capacity is costly. To some degree network operators are prepared to make these investments because they realise that, if their networks were to fail frequently, then they would lose customer confidence and soon after their business. However, they might not be prepared to invest as much as governments would like. This is because the incentive for the operator is to make their network just reliable enough that subscribers do not migrate to other networks on account of reliability. For governments the desired outcome is for networks to be reliable enough to provide communications in times of crisis. The two motivations are completely different and may correspond to different requirements. If government requirements are much more stringent, it may be the case that subsidies or other incentives need to be provided to network operators.

21.5 Towards universal service

Typically, as services such as water and electricity became a core part of our daily lives, governments decided that they were essential utilities and mandated universal provision to all households (although alternative approaches such as local generators are allowed for very remote dwellings). Fixed-line telecommunications to the home was declared such a service in most developed countries many decades ago and there is now a universal service obligation in most countries. To date, few countries have extended this requirement for universality to broadband fixed communications or to mobile communications, but debate about whether this should occur is now happening. Indeed, in some countries broadband and mobile discussions overlap since wireless networks are seen as one key element of the solution of providing broadband service to the home.

Regulated utilities with universal service requirements are normally monopolies. In most countries there is only a single water pipe going to the home, a single power cable and so on. Competition can occur within the system – for example, in electricity there are competitive generators and competitive supply companies that can administer the supply over the single network. It is difficult to impose universality in a truly competitive market because this requires regulation of each of the competitors, which can be costly and distorts competition. For example, as has already been mentioned, if government subsidy were to be provided to mobile operators in order for them to extend their coverage, then this would need to be provided three or four times over to all the operators, possibly with the level of subsidy for each depending on the state of its existing network. Strict rules would need to be in place to cover eventualities such as network mergers. Most would see such duplicated funding as inefficient and the resulting restrictions as overbearing and inflexible.

If governments do decide that mobile coverage is insufficient to meet the needs of their citizens and that they should intervene, then the net result is likely to be fewer networks, with the radio network being shared among multiple operators. There may only be a single extensive 'primary' network in the country, which has been subsidised to provide widespread coverage and is used by all the mobile operators, perhaps via a roaming agreement, when outside the coverage of each of their individual networks. Some operators may decide that owning their own network no longer generates any competitive advantage for them and may fold it into the primary network. Some countries may go as far as ending up with a single network that all the operators – which have now become mobile virtual network operators (MVNOs) – use to provide service.

Such networks may have advantageous frequency bands reserved for them, since the cost of providing universal coverage is less at lower frequencies. They may also have wider frequency bands made available, since this would facilitate extensive broadband provision using technologies such as LTE. Already in some countries there is discussion of the proposal that the 'digital dividend' spectrum liberated when analogue TV is switched off, which is at about 800 MHz, should be set aside for operators willing to provide extensive rural broadband services.

Each country has a different starting point in terms of the number of operators and the level of coverage that they provide, and each may have a different desired goal depending on their inclination towards encouraging competition versus provision of services to citizens. Hence, the paths followed and the outcomes may differ. In general, it seems likely that mobile networks will becoming increasingly like utilities, regulated to ensure that widespread access is possible and tending towards fewer networks, with competition occurring at the service level. The reduction in competition may result in better coverage but less innovation, with less inclination to roll out new technologies. The annoying not-spot may yet be the mechanism by which mobile networks become just another utility.

22

Summary

22.1 Key technologies

We started by examining all of the known new technologies currently 'on the wireless horizon' or, in some cases, much closer to implementation.

We looked at fourth-generation cellular systems and noted that they might bring some advantages in terms of both higher data rates and more efficient use of spectrum. With new spectrum becoming available to cellular operators in bands such as UHF (between 500 and 800 MHz), 2.6 GHz and 3.4 GHz, there is an inclination to use this for a new generation of technology rather than deploying more 3G, and this additional spectrum alone will provide much additional capacity. However, with cellular capacity rapidly being consumed by data, end users might not notice a substantial difference on going from 3G to 4G – instead the technology may be more about reducing the operator's cost base. A major question mark over 4G is the extent to which MIMO can bring benefits in real deployments; if it does not, then many of the promised gains of 4G will not prove to be real.

Femtocells are a topic of much current interest. We are certain that there will be small cells in the home – indeed, there already are WiFi hotspots in many. What is less clear is whether femtocells will be deployed in addition to WiFi. Much of this depends on the business models of the cellular operators; already different operators are deploying different models. Perhaps, as with the handset market, different operators and different homes will deploy differing solutions depending on their preferences and we will see a mixed model. The inclusion of small cells into a converged wireless communications system will change much; whether these are femtocells or WiFi matters far less.

Mesh networks are seen by many as the future of communications, mostly because they are self-organising and can be built up independently of operators.

However, they have many disadvantages, which mean that they are ill-suited for real-time communications or communications involving moving terminals. Mesh does have a role to play in a few niche applications, key amongst which appears to be linking sensor networks. Other applications might include building communications networks where infrastructure has failed and communicating within closed groups where all the members are in the same vicinity – for example, emergency services or military teams.

Software-defined radio systems have been the subject of research and some limited deployment for more than a decade. The idea of a common hardware platform that can be defined and controlled by software owes much to PCs and is compelling for many reasons, including cost and flexibility. In practice, fully software-defined radios are impractical and even highly flexible ones are overly expensive. Software definition makes more sense for base stations, which have a longer life and where power supplies are less of a problem. Indeed, some software-defined base-station platforms do exist. Software-defined devices to date have simply been an alternative way to make wireless devices, but have not had materially different functionality. At present it is hard to see this changing.

We noted that cognitive devices had the potential to access a substantial amount of spectrum that was unused. However, this comes at the cost of increased complexity and it is not clear that there is any compelling need for more capacity, although this may provide a useful alternative if WiFi congestion increases in the future. We surmised that cognitive radio was a 'solution in search of a problem' and that, until a problem emerged, cognitive access was unlikely to become widely deployed. However, while it still has the backing of many major companies anything is possible.

Codecs can improve the functionality of wireless systems by reducing the bit rate needed for audio and video communications and hence effectively enhancing network capacity. However, audio codecs appear to have reached a point at which further improvements are minor. Video codecs exhibit greater room for improvement, but this is tempered by the difficulty of introducing new codecs and the ever-increasing requirements for higher-definition broadcasting. It seems highly unlikely that advances in codecs will result in reduced data-rate requirements in the future.

Devices make a big difference. As the iPhone has shown, a better user interface can result in large increases in wireless traffic and a wide range of new applications. Further improvements in screens, keyboards and other input/output devices could bring about similar step changes in the uses to which wireless devices are put. Foldable or rollable screens have been promised for many years and demonstrated as prototypes, but remain very difficult to manufacture in volume. Projection devices are now available in some phones, but they increase the bulk of the phone and a static environment with a suitable wall or screen is required for viewing. Many different keyboard arrangements have been tried, but none is as good as a full-size PC keyboard. Voice recognition could provide an alternative and is steadily improving, but will take decades to become near-perfect. Battery life will remain a constraint, but storage will not. Phones will continue to gain more functionality, which could lead to many new uses of phones as probe devices or similar.

Overall, when considering technologies, we see nothing on the horizon that will revolutionise wireless. For the user, steady improvements in devices will facilitate additional uses that are already feasible today but not sufficiently simple. For the operator, 4G may reduce network costs, while integrating small home cells into networks will become critically important. For the manufacturer, software-defined elements of wireless devices may grow, changing the balance of device design more towards software. For the unlicensed user, cognitive devices offer possibilities for new applications or for congestion relief, although for many these might not be necessary since WiFi will continue to meet their needs. This all adds up to solid incremental improvements but not a revolution.

22.2 Key solutions

We started our consideration of solutions with a look at users and how their behaviour has changed. One of the key messages was that the industry frequently mis-forecast what users wanted, often delivering services that were unsuccessful (such as video-calling) and being surprised by the success in other areas such as ring-tones and texting. The chapter concluded that the mobile phone would become ever more personal, being used for social communications, entertainment and distraction much more than for 'valuable' communications. This would continue to be difficult for the industry to understand, with the result that value would often be captured by others such as Apple, social-networking websites or publishers.

Sensor networks offer solutions to a number of problems. They can monitor health, better control buildings to reduce energy consumption or monitor the environment more widely. The ideal of tiny sensors that can be 'scattered' and then forgotten seems unlikely to be realised, since the problems of powering them are just too great. However, sensors in building and worn by individuals are much more likely and, indeed, are already emerging. These can make use of a number of developed standards and a range of network architectures depending on their requirements. Sensors will require a mix of small home cells and/or mesh networks to interconnect them.

Home cells will be a core part of a hierarchical network architecture. This will become increasingly critical as data volumes rise on cellular networks. As this develops, devices will become 'always best connected', although operator business models may act to prevent this in the short to medium term. In the longer term we can expect operators to adapt to the hierarchical model.

Healthcare is an obvious area where wireless can bring benefits. Sensors can check on the condition of individuals and sound alerts, making it possible for some to live longer in their own homes. Wireless in the hospital can bring efficiency benefits. Wireless healthcare broadly builds on sensors, mesh networks and small cells in the home or hospital – all of which are available today. However, the pace of change in healthcare is slow, for a number of reasons, including funding issues, risk aversion and the difficulty of developing standards.

The prognosis for transport is similar to that for healthcare. Wireless can bring benefits, including better route guidance, improvements in vehicle safety brought about by avoiding collisions and sharing information, luggage tagging at airports and passenger communications from planes and ships. However, any solution that requires collaboration of manufacturers, governments and others is difficult and slow. Such solutions are likely to be overtaken by standalone systems that can be fitted (or retro-fitted) without any need for widespread agreement, even if the functionality is not quite as good.

Entertainment has been one of the largest users of wireless communications and this will continue to be the case. The biggest demand comes from video content and a key uncertainty in the future is whether users will predominantly 'side load' this content from a home server to their device or will stream content to their device when out of the home. In practice, the costs of the latter suggest that it will be adopted only for material that has a high value when live (e.g. news and sports events) or when the device is in a hotspot and so able to download at a much lower cost.

Assisted-living services are critical for an ageing society and wireless can play a role. Much of this is the same as for healthcare – sensor networks coupled with the provision of information can allow people to live in their own homes for longer. There may also be a role in providing engagement services to allow people to remain active members of society even if they cannot get out so much and allowing some to telework so that they can continue to contribute. As with healthcare and transport, the issues here are not so much a lack of technology but issues of funding, conservatism and the difficulty in agreeing standards.

Overall the picture is one of wireless having adequate technology but being held back by the complexities of introducing new services into environments such as transport and healthcare. Where little agreement is needed, rapid advances can be made – the iPhone is a classic example of this. Where agreement of manufacturers and government bodies is needed, progress is typically very slow – so much so that the optimal solutions may be side-stepped by alternatives that are less optimal but simpler to implement.

22.3 The journey has to be worthwhile

This book is not intended to be a comprehensive prediction of the future – other texts that attempt to do this are available. Instead, it looked at individual services and solutions that are on the wireless horizon, without trying to draw too strong a linkage across different technologies and different services.

One theme emerges. Much is possible, but often the need for the advance is not compelling or the institutional and competitive arrangements are such that new services are very difficult to introduce. Many forecasts for wireless futures have proved hopelessly optimistic not because they over-estimate the advances in technology (although some do this as well) but because they under-estimate the institutional difficulties or fail to understand what it is that users really want from mobile devices. The visibility of something new on the horizon is no guarantee that the journey to find it will be worthwhile.

Index

Printed in the United States
By Bookmasters